An Ethnography of Global Environmentalism

Based on nine years of research, this is the first book to offer an in-depth ethnographic study of a transnational environmentalist federation and of activists' work and life. The book presents an account of the daily life and the ethical strivings of environmental activist members of Friends of the Earth International (FoEI), exploring how a transnational federation is constituted and maintained, and how different people strive to work together in their hope of contributing to the creation of "a better future for the globe." In the context of FoEI, a great diversity of environmentalisms from around the world are negotiated, discussed and evolve in relation to the experiences of the different cultures, ecosystems and human situations that the activists bring with them to the federation. Key to the global scope of this project is the analysis of FoEI experiments in models for intercultural and inclusive decision-making. The provisional results of FoEI's ongoing experiments in this area offer a glimpse of how different notions of the environment, and being an environmentalist, can come to work together without subsuming alterity.

Caroline Gatt is a Research Fellow in the Department of Anthropology at the University of Aberdeen. Her research interests include environmentalism, post-relativist approaches to human diversity, laboratory theatre and design anthropology.

Routledge Studies in Anthropology

www.routledge.com/Routledge-Studies-in-Anthropology/
book-series/SE0724

An Ethnography of Global Environmentalism

Becoming Friends of the Earth

Caroline Gatt

Routledge
Taylor & Francis Group

LONDON AND NEW YORK

First published 2018 by Routledge

2 Park Square, Milton Park, Abingdon, Oxfordshire OX14 4RN
52 Vanderbilt Avenue, New York, NY 10017

Routledge is an imprint of the Taylor & Francis Group, an informa business

First issued in paperback 2020

Copyright © 2018 Taylor & Francis

The right of Caroline Gatt to be identified as author of this work has
been asserted by her in accordance with sections 77 and 78 of the
Copyright, Designs and Patents Act 1988.

All rights reserved. No part of this book may be reprinted or
reproduced or utilised in any form or by any electronic, mechanical,
or other means, now known or hereafter invented, including
photocopying and recording, or in any information storage or
retrieval system, without permission in writing from the publishers.

Notice:
Product or corporate names may be trademarks or registered
trademarks, and are used only for identification and explanation
without intent to infringe.

Library of Congress Cataloging-in-Publication Data
A catalog record for this book has been requested

ISBN: 978-0-415-71762-5 (hbk)
ISBN: 978-0-367-59420-6 (pbk)

Typeset in Sabon
by Apex CoVantage, LLC

In memory of Julian Manduca

One of the pioneers of the Maltese environmental movement, long time co-ordinator of Moviment ghall-Ambjent

—Friends of the Earth (Malta).

Contents

Figures

Acknowledgements

To make this book possible so many people have been generous with their time, energy and trust. I would like to thank everyone who has contributed to this work along the way. In particular, I would like to thank my family, foremost my late parents who always encouraged curiosity, also my brother Michael who supported me throughout my research in more ways than can be mentioned. Heartfelt thanks are due to my cousin, Cathy Farrugia, for proofreading an earlier version of this manuscript at short notice, and for her uplifting comments.

I am indebted to all the Friends of the Earth activists whose unrelenting work, often with the smallest resources, is a labour of hope. Without their work and lives, not only would there have been no study but, I am certain, a different world altogether. In Malta, special thanks to Martin Galea de Giovanni, a dear friend, who trusted me and helped me, even during writing up when I disappeared for years at a time. In Brazil, I would like to thank Nely Blauth, who offered me the warmth of family and took care of me when I was ill. Thanks to Lucia Schild Ortiz, Kathia Vasconcellos and Daniele Sallaberry for making me feel at home in Porto Alegre and being enthusiastic about my research project. Special thanks to my flat mate Anne Wilson and my friends Carla Villanova and Elisangela Soldatelli for the many long conversations. In Amsterdam, Ann Doherty for welcoming me and championing my fieldwork in the IS, for offering me a home in my first month of fieldwork and for her feedback on the chapters in this book. Thanks to Mae Ocampo and Wieke Wagenar, who supported me and my work in the Secretariat, even when it was not clear to anybody what I was doing there. To Maria Enriqueta Homrich, from FoE Brazil, who helped me build trust with others at the FoE IS in Amsterdam. Debra Wilson, Sisi Nutt, Ana Santana and Ginting Longgena, thank you for your friendship and for sharing your thoughts as well as our experiences. Thanks to Tchitula Gisah and Elrieke Muller for giving me a home outside of the office of FoEI, and giving me something else to think and talk about apart from FoEI—probably to the benefit of those around me. Some names and identifying details have been changed to protect the privacy of individuals.

At the Department of Anthropology in Aberdeen, where the project started, I would especially like to thank Tim Ingold, Arnar Arnason, David

Acknowledgements

Anderson and Andrew Whitehouse. Throughout the process of the research they have given me inspiration, clear guidance and endless patience in working through my drafts. I would also like to thank Nancy Wachowich, Alison Brown, Amber Lincoln, Sophie Elixhauser, Peter Loovers and Irena Connon for the rich exchanges about our work, and James Leach for the hours of truly exciting discussions in our exploration of Design Anthropology.

I would like to thank João Pina Cabral and Thomas Hylland Erikson, the participants and the organisers of the Scottish Training in Anthropological Research Courses of 2008 and 2009. Many thanks to Thomas Latour, Neil Washbourne, Cristina Adams, Mark Anthony Falzon, Amiria Salmond, Martin Holbraad, Johan Rasanayagan, Paul Clough, Paul Sant Cassia, David Zammit, Elise Pisani, the late Jeremy Boissevain, Jonas Tinius, Brian Doherty, Rolland Munro and Nayanika Mathur. Thanks to Arturo Escobar for his encouragement in developing this work into a book. Jon Mair and Liana Chua, thanks for your encouragement, guidance and generous collegiality, I treasure it. I would also like to thank Matei Candea who provided guidance and mentorship while I was preparing the manuscript.

I was able to carry out the research thanks to the support of the University of Aberdeen's Sixth Century Studentship and Small Grants Fund. My fieldwork period in Brazil was made possible thanks to an affiliation to the Universidade Federal de Sao Paolo. A thousand thanks to Patrick Mifsud, without whose help I would not have got a visa for Brazil. The development of the manuscript was also made possible thanks to the Sociological Review Fellowship that I was awarded from 2011–2012.

From 2013 I have worked on the project 'Knowing from the Inside', funded by the European Research Council, at the University of Aberdeen. There I have worked with and learned from all my colleagues. Special thanks to Tim Ingold, Marc Higgin, Germain Meulemans, Francesca Marin, Alyson Millar, Ester Gisbert Alemany, Camille Sineau and Valeria Lembo.

I presented parts of this manuscript in earlier stages at the University of Aberdeen, Cambridge, Oxford, St Andrews, UCL, Edinburgh, and meetings of the American Anthropological Association, Association of Social Anthropologists and European Association of Social Anthropologists.

Chapter 2 was previously published in an altered form as "Vectors, direction of attention and unprotected backs: Re-specifying relations in anthropology", *Anthropological Theory*, Vol 13 Issue 4 (2013): and "Enlivening the supra-personal entity: vectors at work in a transnational environmentalist federation", *Anthropology in Action* Vol 20 Issue 2 (2013). Chapter 3 was published as "Emplacement in multi-sited practice/theory", in Falzon M.A. (Ed.) *Multi-sited Ethnography: Theory, Praxis and Locality in Contemporary Research* (UK: Ashgate Publishing Ltd.) (2009).

At Routledge, I would greatly like to thank Max Novick for years of understanding and guidance, Katherine Ong and Marc Stratton for all their support and help with the nitty-gritty details in preparing the manuscript.

Most of all I want to thank Richard Muscat and my daughter Mariuccia Muscat, for being more than a little patient throughout the process of writing and preparing the manuscript. I am indebted to Richard for the many conversations during which he helped my ideas crystallise and for the diagrams that are found in this book.

Sincere thanks.

Abbreviations

AD	Alternattiva Demokratica
ADFG	Ação Democrática Feminina Gaúcha (Gaucha Democratic Female Action)
AGM	Annual General Meeting
ANT	Actor Network Theory
Anti-GE	Anti-Genetic Engineering
ATALC	Friends of the Earth Latin America and the Caribbean
BGM	Biennial General Meeting
BINGO	Big International non-Governmental Organization
CAN	Climate Action Network
CV	Curriculum vitae
EGM	Extraordinary General Meeting
ENGO	Environmental Non-Governmental Organisation
EU	European Union
ExCom	Executive Committee
FoE	Friends of the Earth
FoE EWNI	Friends of the Earth England, Wales and Northern Ireland
FoEE	Friends of the Earth Europe
FoEI	Friends of the Earth International
GM	General Meeting
GMO	Genetically Modified Organisms
GT	Grupo de Trabalho (Working group)
ICT	Information and Communication Technologies
IIRSA	Initiative for the Integration of the Regional Infrastructure (South America)
IM	Instant Messaging
IMF	International Monetary Fund
INGO	International Non-Governmental Organisation
IP	Internet Protocol
IS	International Secretariat (FoEI)
MEDNET	FoEI Mediterranean Network
MEP	Member of the European Parliament
MEPA	Malta Environment and Planning Authority

MITTS	Malta Information Technology and Training Services
MSF	Membership Support Fund
NAT Brasil	Nucleo Amigos da Terra/Brasil—Friends of the Earth Brazil
NGO	Non-Governmental Organisation
NT	Nature Trust (Malta)
REACH	Registration, evaluation and authorisation of chemicals
RSF	Religious Society of Friends
SPTT	Strategic Planning Transition Team
STS	Science and Technology Studies
SVPP	Strategic Vision and Planning Process
TNC	Transnational Corporation
UFRGS	Universidade Federal do Rio Grande do Sul
UN	United Nations
UNFCCC	United Nations Framework Convention on Climate Change
WB	World Bank
WSF	World Social Forum
WTO	World Trade Organisation

1   Introduction

Environmentalism, Globality and Anthropology in our Common World

In 1969 David Brower, then a member of the Sierra Club, a leading US environmentalist organisation, despaired of the organisation's conservationist approach and refusal to address questions of justice. He left his position in the organisation and set up Friends of the Earth (FoE) in the United States. Two years later, in 1971, together with activists from England, France and Sweden, Brower created an international federation that they called Friends of the Earth International (FoEI). Today, more than forty years later, the official website describes FoEI as 'the world's largest grassroots environmental network, uniting 75 national member groups and some 5,000 local activist groups on every continent. With over 2 million members and supporters around the world'.[1] FoEI differs from many other environmentalist federations because, rather than being solely concerned with nature conservation, the federation is equally concerned with social justice. A number of academics (Doherty 2006; Timmer 2007), as well as many FoEI activists, compare FoEI to Greenpeace. In this comparison, the two organisations differ primarily due to FoEI's emphasis on a decentralised organisational structure. Indeed, FoEI expresses this focus on decentralisation by referring to itself as a 'grassroots' federation or network. Although the actual discourse and practice of decentralisation in FoEI is explored throughout this book, the *aim* of decentralisation in FoEI has resulted in significant diversity across the member groups and in the types of environmentalism they bring to the federation. FoEI membership is restricted to one FoE group per country.[2] Membership requirements include independence from religious or ethnic movements, political parties and economic interest groups, and a democratic, non-sexist structure. Member groups range from the primarily nature conservation groups—such as Global 2000 (FoE Austria)—to environmental and human rights advocacy groups—such as Legal Rights and Natural Resources Center, Kasama sa Kalikasan (FoE Philippines). Importantly, each group has its own mission statement, if it has a mission statement at all. The value given to being a 'grassroots' federation is expressed in the importance assigned to these differences. Moreover, many activists maintain and strive for decentralised decision-making as an important quality of their own environmentalism. As a result, FoEI is recognised as one of the main networks of climate justice (Bond 2015).

Above I have offered a description of Friends of the Earth International that resembles the ways FoEI is communicated to broad audiences. Since the rest of the book will destabilise most of the aspects, it is necessary first to offer this picture of FoEI as a point of comparison. This is, after all, the same process I had to undergo in the research; I was offered an official description of FoEI and along the way different aspects became apparent. Friends of the Earth International is, in Eriksen's (2007: 77) terms, a 'truly transnational' organisation; it is an environmentalist federation that includes human rights in 'the environment'. However, the central notions employed both by Eriksen and in FoEI's mission statement—global, transnational, environmentalist and organisation—rather than simply being adjectives, are invitations for ethnographic exploration and analysis. This book is a response to these invitations. It is also a hopeful response. Through this book I hope to join in the efforts being made worldwide by people, some of whom are Friends of the Earth activists, to work towards a more sustainable, and equitable, shared world. The particular approach of anthropology is to join in that endeavour with critical ethnographic attention to the constitutive forces at play in the contexts we work in, but also to offer speculative imaginings that, whether deliberately or not, have political and constitutive effects of their own.

For a long time anthropologists argued that the method of long-term fieldwork in locations distant from places they were familiar with was essential in developing that particular attention considered 'ethnographic'. One key element in this ethnographic attentiveness, but by no means the only one, is the ability to become aware of one's assumptions about the world and how life is, how it has come to be that way, and how it should be lived. Carrying out fieldwork in distant, unfamiliar places was one way to bring such assumptions into awareness. Confronted with radically different ways of life, those anthropologists educated in formal Western educational systems found all sorts of taken-for-granted things like religion, economics, family, history, politics and, of course, nature, if recognised at all, to be construed quite differently. Broadly speaking, this highlights anthropology's commitment to developing theory, drawn from empirical findings, that at least does not contradict the experience of informants (Coleman and Collins 2006; Amit 2000). This is an amended position from Malinowski's programmatic statement of trying to understand 'the native's point of view' (1922: 25), or in Geertz's formulation 'the actor's point of view' (1973). With this fundamentally important qualification, fieldwork can be thought of as a method of defamiliarisation through which, by trying to understand a way of life different to one's own, one could not only come to appreciate the great diversity of human life but also to question one's own way of life, and to practice a form of cultural critique (Marcus and Fischer 1999).

It took quite a tempestuous debate in anthropology, still far from resolved, to acknowledge that even with very rigorous preparation, anthropologists could still reproduce their own assumptions even when confronted with ways of life apparently radically different from their own. Anthropological accounts risked 'explaining away' (Strathern 1990) the 'explanations' of the world of the people with whom anthropologists worked, the very people whose 'point of view' they were trying to understand. In this way, they could end up not really questioning their own assumptions at all. This realisation leads Viveiros de Castro (1998) to argue for the need to 'take seriously' accounts encountered in fieldwork, and to recognise the effects of epistemological colonialism (Viveiros de Castro et al. 2014). Bearing in mind Viveiros de Castro's specific proposals and what he means by taking things seriously, the experiences, struggles and hopes through which FoEI activists live raise important questions about politics, ethics and ontology. The questions the work and lives of FoEI activists raise relate to the ways anthropologists and FoEI activists imagine belonging: their own belonging and the world they belong to as well as the world they work towards.

Globality and Ontological Politics

Since the 1980s, more and more anthropologists have been doing fieldwork 'at home', challenging the received distinction between sociology as the study of the complex and anthropology as the study of the distant and small-scale societies of other people. Doing fieldwork 'at home' once again raises the methodological question of familiarity. Would the ethnographer be able to develop the distance required for a critical analysis if they were 'too close' to the context of study? One thing to note was that in the 1980s a lot of what was called anthropology at home was not actually in places or contexts that the anthropologists carrying out the studies were familiar with at all. Although the anthropologist may not have left their native nation-states to do their fieldwork, it was nevertheless carried out in remote communities or among groups from very different socio-economic backgrounds (Strathern 1987). Only few works actually include university life (Bourdieu 1984; Peacock 2013) or the anthropologist's own childhood (ex Okely1992) as part of the explicit context of study. Often reflexivity was considered to be 'navel gazing', and autoethnography was barely mentioned in major anthropology journals—although nowadays gaining more attention, this is mostly in other disciplines such as education and art.

My first job after graduating from university with a BA degree in Anthropology was as a project manager for FoE Malta. For me, as for many of the activists I worked with in this protean thing we referred to as FoEI, the world was experienced as a single place. This sense of cosmopolitan belonging is far from unique to FoEI activists. It is held by many environmentalists, for whom it is often associated with the Gaia hypothesis (Milton 2002: 31). Not only do many environmentalists experience the world as an unbounded and differentiated yet commonly shared place, they also feel it *should* be thought about and treated in this way—that is, if we are to respond well as humans to the various crises of weather, water, food, pollution, safety and loss of habitat that we and the non-humans in our world are experiencing. In fact, in the last few decades, what are marked as *global* crises have become primary concerns for people and institutions (Beck 1992). Environmentalists have played an important role in bringing these concerns to the forefront (Beck 1992; Giddens 1994; Berglund 1998; Castells 1996; Boissevain and Gatt 2011).

Immediately, anthropologists' alarm bells began ringing, and quite rightly. Anthropological research has revealed how environmentalism has disenfranchised indigenous groups, amongst others (Anderson and Berglund 2003: 5, 10, Adams 2003). The negative effects on Inuit livelihoods brought about by Greenpeace's anti-seal hunting campaigns are well documented, to the point that Greenpeace was compelled to change its position on seal hunting and now supports Inuit hunting practices.[3] Even where environmentalists do not have as much influence as Greenpeace, they may still be the powerful elites of the future, as Anderson and Berglund (2003) warned more than ten years ago. This is precisely the critique levelled at large transnational conservationist NGOs today (Larsen 2016), and it is particularly important for two reasons. The first lies in an assumption of scale built into the environmentalist drive to consider the world as a commonly shared place, in which events happening locally are defined as having global import. Anna Tsing writes that "[i]n these times of heightened attention to the space and scale of human undertakings, economic projects cannot limit themselves to conjuring at different scales—they must conjure the scales themselves. In this sense, a project that makes us imagine globality in order to see how it might succeed is one kind of 'scale-making project'" (Tsing 2005: 57). When activists conjure this global scale, we also need to ask which scales are being discounted? What other topologies may be obscured by the appeal to the global? The second reason, related to the question of scale, lies in assumptions about difference: how is difference recognised, understood, brought into play in the work and encounters that are counted as 'FoEI'? What work is alterity made to do in the strategies and tactics involved in the transnational world of NGOs and activism?

Anthropological analysis and questioning has always meant formulating ontological proposals, even if entirely implicitly. By offering an explanation or description, a writer is inevitably presenting one possible ontology rather than another, no matter how implicitly or unintentionally. What we have seen more recently is an *explicit* debate over ontological matters. This may be part of a widespread concern with ontology that exceeds the disciplinary boundaries of anthropology. Mario Blaser suggests that "[t]he present moment can be most fruitfully understood as marked by the increasingly visible and generalised ontological conflicts that are associated with the struggle to shape the global age as an alternative to, rather than a continuation of, modernity" (2010: 1). Ontological conflicts are central to the times because they reveal that there already exist alternatives to modernity and "because they force modernity to reshape itself in order to deal with radical difference" (Escobar 2008: 2). To adopt Blaser's terminology, the stories

FoEI activists tell about the global environment participate in these onto-logical conflicts, and in making globalization itself, in a very tangible way. This making of globality is carried out a great deal through discourse but is definitely not limited to it. FoEI activists' sense of globality participates in ways that I will explore in this book, in making globalization in the onto-logical sense that their actions and understandings of and in the world have effects in generating that very world.

The global scale that FoEI activists have been campaigning about for the last forty years has now gained immense traction within scholarly circles in the debates around the so-called Anthropocene. Introduced by the atmos-pheric chemist Paul Crutzen, the Anthropocene is supposed to denote a new geological era in which human actions, more than other forces, define the global environment. The Anthropocene describes an "environmental glo-balism" in which "it is virtually impossible to disentangle the social and the natural" (White 1999: 979). However, the term, and the ontological assumptions it carries, risks undermining the onto-political work of FoEI.

Scholars have criticised its implied anthropocentrism, but as Sayre notes, these critics have nevertheless continued to use the term and "seemingly grant it de facto conceptual purchase" (Sayre 2012: 61–62).

> The consequences that come along with the Anthropocene thesis, namely, that if humans have unwittingly become a telluric force reshap-ing the planet, how are they going to consciously steer away from its pernicious consequences? This raises the question of politics par excel-lence, that is, how can be constituted the collective subject that will address this challenge and in which ways will do it?
>
> (Blaser nd)

The attention to difference that emerges from my work with FoEI requires a focus on politics and power, specifically due to the questions of *onto-logical* politics that FoEI activists have long been engaged in, although in FoEI terminology this is referred to as 'cultural justice'. A focus on poli-tics and power is not only relevant but necessary for a critical exploration, since so much of environmentalist discourse is pitched in depoliticised terms (Latour 2003; Little 1999; Fischer 1997). In this discourse activist groups, especially non-governmental organisations (NGOs), are regarded as being benevolent and non-political, especially due to their independence of the state and profit-making economic activity (Fischer 1997). Other strategies mobilised by environmentalist organisations depoliticise the issues at hand by, for instance, defining matters as 'technical' (Müller 2011, 2013), or by justifying their actions in terms of using necessary 'evil means' for 'good ends' (Larsen 2016). Environmentalists in the NGO context strive for legiti-macy as authorities on the management of land, resources and morality, based on their scientific expertise and benevolent intentions (Gatt 2001; Lis-ter 2003; Choy 2005). However, it has also been found that environmental degradation is most acute where inhabitants belong to subordinate, politically muted social groups within a society (Bullard 1994). The work of activists and the situations in which activism arises are inevitably underpinned by power struggles, but as I will discuss in more detail in a subsequent section, exploring the questions about power that the specifically 'global' experiences and striving of FoEI activists raise requires that we think about the forms, and rhetorical strategies, that our scholarly arguments take.

Here it is necessary to return to the question of familiarity in relation to ontological politics. How the anthropologist defines their own belonging in relation to the context of fieldwork affects the form their scholarly argument will take: Closeness or familiarity can also make evident the great degree of variation in the 'West', a term and imaginary which often functions as a foil for anthropological analyses of 'the rest' (Chua and Mathur 2017). Both Orientalism and Occidentalism take shape by eliding the particularities and heterogeneity of the ways of life within the territories supposedly in question (*ibid*). Once we begin to pay attention to the infinite variation in every circumstance, the rhetorical devices of contrast (us and them, the West and the rest), and the strawmen they construct, lose their persuasiveness.

Recalling Viveiros de Castro's injunction to take others seriously: focusing on the experiences of FoEI activists and their questioning of the global means also taking seriously thoughts and concepts that originate in the 'West'. Viveiros de Castro's own formulation, dependent as it is on opposing different 'worlds' (Graeber 2015), makes 'taking seriously' hard to apply to one's own concepts, at least without qualification (Candea 2010). Not only is it necessary to acknowledge the actual effect that concepts originating from Europe or the United States can have on the rest of the world, and in so doing to recognise coloniality (Blaser 2010; Escobar 2008) and global inequalities (McGill 2016), but also to take serious ethnographic interest in otherwise anthropologically suspect things such as 'democracy', 'liberalism', 'individualism' and 'universals' reveals that none of these concepts and discourses, nor the ways of life in which they participate, are or ever were hermetically sealed, 'pure' or homogeneous. Back and forth influences across multiple places have been present throughout history. In fact, in trying to make this argument, I find myself in a similar position as FoEI activists themselves, striving to 'take seriously' the ontological politics of other FoEI activists from all around the world, without having one's own voice drowned out.[4]

The particular challenge FoEI activists face is how to develop joint policies for FoEI, a federation that not only includes very diverse forms of environmentalism and understandings of 'the environment' and 'the world' but, as a matter of principle, also promotes that diversity. The challenge is always present in any FoEI or FoE group campaign. In their campaigns, activists endeavour to convince their various target audiences, such as the United Nations (UN), the European Union (EU), national governments, shareholders of multinational corporations, a member group's national public, and so on, that their policies and their mediation will *improve* the world shared by all humans and all non-humans alike. Explaining the difference between environmentalist and non-environmentalist ideas, as Argyrou (2005: 5) does, by appeal to their historical and cultural specificity, or by putting things in their historical and cultural contexts, can account neither for conflict or disagreement (Asad 1979), nor for how we can become aware of and learn about difference in the first place (Ingold 1993).[5]

The activists I worked with are convinced that different human ways of life have actual effects in a single common world that we all share. Those ways of life often result in oppression or destruction of other people and non-human environments. Activists are highly attentive to ontological politics also in the sense that they take seriously each other's relationships with non-humans. In this book, I explore different dimensions of power in the daily lives and work of FoEI activists, in their dealings with both humans and non-humans: How can one simultaneously take into account these different dimensions (material agency, personal agency, impersonal structures, symbolic and institutional power, etc.) within a shared world? The question is how to conceptualise different forms of power simultaneously, without explaining one form in terms of another, and thereby producing a reductionist account.

What emerges from my exploration of FoEI practices and experiments in politics and policy-making is that activists participate in generating certain types of subject. In this book, I follow activists' efforts to lead ethical lives. We see how activists strive for an ethical and sustainable life, and how the processes in which they participate in turn generate certain types of ethical subject. We follow not only the activists' own ethical self-formation but the work that goes into collective ethical formation.

In the remaining sections of this introduction I explore in more detail the conceptual underpinnings of the three themes I have introduced so far: globality, ethical formation, and ontological politics, with a focus on how to 'take seriously' such politics within a common world. I hope to offer the reader a particular stance—conceptual, ontological and attentional— through which to receive the ethnography that follows, and to direct the reader's attention to current debates in the literature in a way that will make evident my choices of what I explore and what I describe in the following chapters.

Environmentalism and 'the Global'

Environmentalists, including FoEI activists, are well known for having argued that environmental disasters do not respect national boundaries. Different forms of environmental degradation include global threats: acid rain, river pollution, depletion of fish stocks, marine pollution and most recently climate change. All are transnational problems. Thanks to the ideas and practices of activists, social theorists have identified environmentalism as both a product as well as a force of globalisation (Giddens 2007; Harvey 2000; Castells 1996, 1997; Beck 1992, 2007). In anthropology, especially, globalisation has been a highly contested matter. Is it as new a phenomenon as is claimed? There have, in fact, been many world systems prior to the current period, although certain aspects of current trends may be unprecedented (Eriksen 2003, 2007; Hannerz 1996). Is there such a thing as 'the global'? Some argue that there definitely are global ideologies, for instance Hannerz's (1996) 'global ecumene', and Tsing's (2002) 'ideologies of scale' and Mosse's (2005) 'transnational episteme'. Others argue that any notion of the global is necessarily localised, an embodied position, and that no abstract global point of view can exist (Haraway 1991; Ingold 1993). Inda and Rosaldo (2002) argue that any cultural notions or practices that travel and can be understood as global must first be disembedded from one locale or territory and then, simultaneously, re-embedded in another territory: what they call 'de/reterritorialsation'. Milton, on the other hand, imagines environmentalist discourse to be highly mobile and free-floating (1993). Robertson (1995) proposes the term 'glocal' to emphasise that what is considered 'local', at times, becomes more valued as such because of 'global' pressures.

Most literature in anthropology explores world systems, globalisation and transnationalism.[6] These studies emphasise how the universal and the global are everywhere, in specific ways, and affect how people interact and lead their lives (Tsing 2005; Choy 2005).

> While a world view is always a view held by someone, global as in 'global culture' refers not to the view but to the world. So to speak of the global is less to describe a world view than to hold one oneself.
>
> Marilyn Strathern (quoted in Berglund 1998: 179)

In 'Globes and Spheres' Tim Ingold argues that a 'global' perspective necessarily implies not only taking a distance from the world—physically possible only in a spacecraft—but also its objectification as a separate 'thing' to be analysed and appropriated (2000: 214–216). In Euro-American educational systems' maps, globes and nowadays satellite imagery are the means by which a 'global' perspective is transmitted to the overwhelming majority of people (barring astronauts) who do not have access to a direct perception of the world as a globe. Ingold opposes this 'world view' to a spherical lifeworld. Although people with a global world view may experience a sense of alienation, it is not possible for them to step outside of the spherical lifeworld. Even those few astronauts who have perceived the world directly as a globe have nonetheless lived that happening at the centre of their sphere of experience. Their relation to the elements all around them continued to constitute their experience, even in space.

Equally, those who have a global world view—whether fostered by maps, globes or (especially in the case of some FoEI activists) regular aeroplane journeys—still cannot step out of the sphere of their perception. Similarly to accounts of environmentalism in Beck's *Risk Society (1992)*, among the activists' motivations for being environmentalists is specifically a feeling of alienation from 'nature' and a deeply felt concern that this could be even further exacerbated. Most FoEI activists simultaneously dwell in the world and hold a global world view of one sort or another. This attention to the global is mainly upheld as a firm resolve that the solution to environmental degradation lies in concerted 'global' action. Although FoEI activists can no more step out of their personal sphere of experience than anyone else, the 'global' perspective forms an important part of their experience and politics.

In documents and campaigns, FoEI claims to be a worldwide federation of grassroots ENGOs (environmental non-governmental organisations). In the context of largely non-ethnographic arguments that environmentalism has shaped and been shaped by the 'forces' of globalisation (Beck 1992; Giddens 1994; Castells 1996), that environmentalism is a further expression of the universalism of modernity (Argyrou 2005), and that the discourses of environmentalism travel (Milton 1993), this study aims to give theoretical discussions of globalisation an ethnographic grounding. In this book, I present an ethnographic exploration of FoEI activists' 'global environmentalism'. Since the global environment is of such central importance to them, the question that runs through the whole book concerns their experience of the global: What does it mean, ethnographically and phenomenologically, when activists claim to engage in global environmentalism? How does the notion of the global environment mesh with a FoEI structure that is based on nation-states? What are the activists' ideologies of scale? What are their experiences of emplacement in relation to the categories of national and global 'levels' of action that they use in their various documents?

Globality and the Anthropocene

Over the years, practices and relationships within FoEI have become increasingly institutional and bureaucratic. Unlike the question of scale in relation to emplacement (local–global), a classical question in anthropology concerns the difference between small-scale and large-scale societies. 'Scale', here, is invoked whenever interactions among individuals are compared to those within the context of a formal institution, bureaucracy or organisation. Personal relations are considered particular and small scale, while institutional or bureaucratic relations are considered abstract and large scale or complex (Strathern 1996; Yarrow 2006). However, approaching institutions and organisations, as I do in this book, from a phenomenological perspective, the question becomes: when persons 'bump up' against institutions, is their experience any less personal simply because an entity is not human-sized?

Previous anthropological approaches to the macro-scale or globalisation have either discarded (Lave 1988) or struggled with (Throop 2010; Nevelling 2010) phenomenology, due to what appears to be its inherently small-scale approach. In Lave's words, phenomenology is theoretically problematic in relation to macro-scale issues because 'phenomenological views consider structure only as epiphenomenal of intersubjective experience' (1988: 193). In this book, I will seek to overcome this objection by focusing on the institutional and organisational aspects of globalisation and transnationalism as actually experienced by activists, while also addressing the gap in the literature on transnational environmentalist organisations: Expanding from the classical question in anthropology: what makes institutions effective? How does FoEI 'hang together' as an international federation? What are the processes by which FoEI activists come to accept that FoEI, and national FoE member-groups, exist and have effect as entities? By what means is the organisation called FoEI constituted and maintained? In what terms does FoEI exist? How can we take seriously both the intersubjective co-creation of meaning and sociality and the effects of history and impersonal structures that cannot be reduced to interpersonal interactions?

Partly in line with the rhetoric of globalisation, a concern has been raised that ethnographic methods, focused as they are on the fine grain of personal relations, might not be appropriate to study such large-scale matters as globalisation or the Anthropocene. However, given the unreachable heights of abstraction to which thinking about globalisation has ascended in social theory (Harvey 2000; Giddens 2007; Castells 1996; Nustad 2003), an ethnographic account of the experience of globality is all the more needed. It offers a reminder that forces such as 'globalisation' always affect the lives of those involved in one way or another. Indeed, if we choose to follow the logic of the Anthropocene, then everyone is now affected by human-caused global changes. Orr, Lansing and Dove (2015: 156) doubt whether a shift to a 'coarse-grained, averaged-out' perspective on planetary-wide matters will offer any more insight than anthropology's person-scaled approach.

In fact, in recent years, big international non-governmental organisations (BINGOs) have attracted criticism precisely for being transnational rather than local and small (Larsen 2016: 24). Naomi Klein claims that "Big green" environmental groups have become part owners of industry under the banner of constructive engagement (cited in Larsen). Holmes (2011, also cited in Larson) traces the relationships among NGO activists, lobbyists and politicians at large UN conferences and concludes that NGOs are no longer local and popular, but global and elitist. Larsen (2016: 25) does wonder whether these critiques are overdone, especially as they seem to appear side by side with government attacks on NGOs. However, whether BINGOs are 'good', as portrayed in their own rhetoric, or corrupted through their co-option by big government and corporations, ethnographic analysis is still needed to explore how such groups negotiate these different scales, between what is considered global and what is considered local. In the case of FoEI, explicit collective work goes into deciding how to address global concerns without forgetting the experiences of people on the ground, or the 'grassroots'. Escobar (2008: 71) notes how in critiques of development, which echo critiques of BINGOs, there has been a certain romanticisation of the local that overlooks the way locales are always involved in power dynamics at various scales.

The opposition between the local and the so-called global carries further assumptions that match with concerns specific to anthropology. If scholars carry out studies of 'global' processes, then this risks granting de facto status to the global—as noted earlier in relation to debates about the Anthropocene. As we have already seen, an anthropological study of global environmentalism risks obscuring its underlying heterogeneity. This critique also attached to the Anthropocene project itself, which seems to attribute responsibility indiscriminately to all humans, when in reality the action that has brought about planetary-wide changes is attributable specifically to heavy industry. For this reason, 'Capitalocene' has been proposed as a better name for the new era. In 2011 Crate wrote that "the industries most responsible for anthropogenic global warming . . . have actively and intentionally stymied effective public response to the problem through a complex web of misinformation, lobbying, intimidation, and public relations." Crate believes that for this reason more attention has been paid in recent years to the ethnographic study of multilateral transnational organisations such as the WTO and UN bodies, rather than to activist organisations. However, at a time when social movements of the Right (Edelman 2001) have gained traction in the United States, in the UK, and increasingly over the last twenty years in other European countries, driving a hopeful democratic process demands arduous and lengthy commitment, especially when, as with FoEI, it includes rigorous attention to ecology and—in their words—'cultural difference', which in FoEI carries political weight. What the Anthropocene enables is a broad discussion, beyond the academy, of how to reconceptualise human relations with world, and this call has come very strongly from social movements, including FoEI (Sayre 2012: 63). According to Sayre (*ibid*: 65), the anthropogenic only finds political expression in the United Nations Framework Convention on Climate Change (UNFCCC). This book will show, to the contrary, that the politics of the anthropogenic, including critiques of the notion itself, can be found in the daily work, negotiations and collective efforts of FoEI activists.

Collective Ethical Formation

In November 2015, the UN held the twenty-first session of the Conference of Parties (COP) in Paris, during which the governments of the 197 signatory countries committed to limiting the world's rising mean temperature to 2°C above pre-industrial levels. These governments prepared plans to reduce their emissions and committed to bringing them into effect by 2020. They also pledged, in the meantime, to 'ratchet up' those plans which, as they stand, are only enough to keep the global temperature increase to between 2.9°C and 3.4°C—that is, if the countries keep their pledge.[7] In preparation for this meeting, a number of BINGOs began mobilising their activists for demonstrations in Paris. Avaaz, an organisation that has more than 41 million 'clicktivists' and backed by Climate Action Network (CAN), decided to organise a single demonstration in Paris on the opening day of the meeting, on November 29, 2015 (Bond 2015: 15). Bond (ibid) focuses on the divergent approaches taken by these NGOs; he quotes the following statement issued by Friends of the Earth International:

> Some organisations, such as Avaaz, are pushing for the big Paris mobilisation to happen on November 29. However, there was an inclusive global process and we collectively decided to present the mobilisations as a package: with decentralised actions on November 28 and 29 and a big mobilisation in Paris on December 12.
>
> Why is Avaaz pushing November 29? Their narrative continues to promote the idea that by simply calling on leaders to act on climate, they will. That's why they only want to mobilise before the COP. Our narrative is that the leaders will not save us until we have built a powerful movement that overcomes the vested interests and forces governments to act. That's why we believe people should have the last word. That is why we have advocated for a centralised moment in Paris on December 12.
>
> FoEI, U.S.-based environmental justice organisations, and others are very concerned with the use of the name and imagery of the 'Peoples' Climate March' (PCM) for our mobilisations this year, as is being spearheaded by Avaaz. Firstly, PCM is too closely associated with some northern organisations. It ignores the mobilisations and important work going on in the rest of the world, especially the south.
>
> Secondly, despite the hard work of thousands of U.S. justice organisations who mobilised people in the communities, Avaaz claimed much of the credit for mobilising 400,000 people. It is likely the same could happen again this year. Avaaz has used its 'success' to dominate climate framing since then, with emails such as '5 months to save the world' or the celebration of G-7 ending fossil fuels only in 2100!! Many of our groups and allies have spent considerable effort rebuilding an alternative and more empowering narrative since Copenhagen that *people* will deliver the transformation, not political leaders. We are concerned that all the climate justice narratives could be undone if Paris is too closely associated with the approach of New York PCM.
>
> Thirdly, we have to be very clear about our peoples' demands for Paris on energy, food, justice and jobs. We have to move beyond marches simply calling for 'climate action', as this is perfectly acceptable to elites since it doesn't challenge their business as usual, doesn't deepen our movements and ultimately lacks the ambition and urgency needed to deliver climate justice.

A number of different contrasts are being drawn here between Avaaz and 'climate justice' networks, such as FoEI. One contrast is between the global North and the global South, a second is between centralised and decentralised actions, and a third—possibly the clearest—between people and elites. Although each of these contrasts draws on particular imaginaries and aims to mobilise different responses in the intended audience, all three convey a specific message about FoEI's representativeness. The statement claims that FoEI carried out 'an inclusive global process' to reach a 'collective' decision; *how* FoEI's positions were reached is essential in this narrative. To FoEI it is important not only to work towards climate justice, with equitable access to food, energy and livelihoods, but also that the work FoEI does is done equitably. The processes of decision-making are of central importance to FoEI activists. To some extent, this comes across already in statements issued by FoEI, such as the one cited by Bond (2015). While a focus on the public texts of such organisations can offer a joined-up narrative of disparate actors, we learn nothing about the processes that animate and lead to social movements' arriving at this or that conclusion. Orr, Lansing and Dove (2015: 157) note how systems' ecology methods can trace the consequences of decisions but not how these decisions are taken. It is difficult to tell in advance whether a law or policy is equitable, except in hindsight, once it has had time to have effects in the world. It is easier to judge in real time whether a decision-making process is being carried out equitably. In the following chapters, I follow the work done by FoEI activists in various situations over nine years of my engagement with Friends of the Earth groups, and pay attention to the multiple strands that make their decision-making processes possible in the form that they are. Rather than paying attention solely to the content of discussion, I explore the way decisions, policies and positions are embedded in a variety of practices and imaginaries.

The Strategic Vision and Planning Process

As I mentioned at the outset, David Brower was the founder of the first Friends of the Earth (FoE) in 1969 in the United States. He invited three other organisations to form FoEI. He felt these organisations shared similar principles and objectives, and that to increase the effectiveness of their activism they would need to organise their international collaboration. Brower has explained that in the initial years any organisation that felt a resonance with FoEI could adopt the name Friends of the Earth, use it in their country and participate in the network. FoE USA, although having initiated the collaboration, had a decision-making power equal to all the other groups.

FoEI was initially modelled upon the representative structure of the United Nations (FoEI Archives, Amsterdam). The aim was to maximise the representative value of their claims: the claims made by FoEI activists to representatives of national governments at international meetings were meant to echo and support campaigns being run in national contexts. Such international contexts included United Nations meetings and other inter-governmental meetings such as those of the International Whaling Commission. Decentralised decision-making was one of the central ideological and political choices that informed the foundation of the network and its subsequent growth. At that time, the activities of FoEI were primarily annual meetings bringing together activists from the national FoE groups, where they exchanged ideas and strategic plans for the upcoming year. Beyond attending these meetings, little else was formally required of the national groups (*ibid*).

During the 1980s, however, the activities and interactions of national FoE groups became more structured and in 1981 an International Secretariat (IS) was established. The Secretariat rotated, operating from a different member group's office each year (FoEI Archives, Amsterdam). Core current activists in the Secretariat interpret this rotation as a result of two factors. First, it was a necessity imposed by the lack of resources of any one group to maintain their national work as well as the Secretariat's work for a lengthy period. Second, it was a deliberate means of keeping in check the power of the Secretariat and preventing its control by any one group. In 1983, when the network had twenty-five members, the Secretariat ceased rotating and settled in Amsterdam with funds from the Dutch government.

In the first fifteen years FoEI activists organised annual meetings, met at international intergovernmental meetings where they were campaigning, visited if they happened to be on holiday in the same place as a FoE office, and corresponded (substantially less than currently) by means of a mailed newsletter, mail and, later, fax. Today, FoEI activists interact considerably more. Biennial General Meetings (BGMs) are statutorily required. During these meetings, an Executive Committee is elected to run the day-to-day affairs of the federation for the upcoming two years. To obtain 'regional balance', this committee is composed of activists from different FoEI regions—that is, approximately an equal number of representatives from each region. The Executive Committee meets at least quarterly. A large international meeting is also organised in the year between Biennial General Meetings (BGMs).

The FoEI regional groupings also have their own annual meetings. Currently, the regions are Africa, Asia Pacific, Latin America and the Caribbean, Europe and North America. Although FoEE was set up in 1985, and Friends of the Earth Latin America and the Caribbean (ATALC) in 2001, the FoEI website and other FoEI-wide activities became sedimented along 'regional' lines during the early stages of a ten-year Strategic Visioning and Planning Process (SVPP) upon which the whole network embarked in 2004. Apart from face-to-face meetings, activists interact by telephone, telephone conferences (Internet telephony services such as Skype have made telephoning even more widespread), email and email groups. 'FoEI All' is an email group for any FoEI activist who asks to be added to the list of addressees. Individual programmes, campaigns and projects each have their own email group. The two member groups FoE Malta and FoE Brazil and the International Secretariat, with which I did fieldwork, have their own internal email groups as well.

In most of the contexts I have mentioned, only FoEI activists participate, but as part of their work they also engage in numerous other fora with activists from other networks and with people who are not activists. The intense communication, extended by communication technology, is, among other things, an expression of the drive to produce communal decisions, campaigns and outcomes. In FoEI communal decision-making is an ideologically driven objective that in some cases is possibly on a par with that of environmental protection. It is recognisable in activists' concern to 'have a mandate' from their group to adopt particular positions and make decisions. This objective is expressed in FoEI's organisational structure and requires considerable effort to enact.

The emphasis on consensus found in international contexts is also evident in the workings of national member groups. For instance, both FoE Malta and NAT coordinators have separately explained their belief that leadership should be about facilitation, ensuring inclusive decision-making and building consensus. The current 'vision and mission' of FoEI is the result of a process called the Strategic Vision and Planning Process (SVPP). The aim of this process was to facilitate exchange of the experiences and ideas of the member groups to establish procedures that work to incorporate their vastly differing ideologies, aims, strategies and so on, into a federation that apart from not breaking up, can be said to work in 'solidarity'. The SVPP was set up because there was a moment in 2002 when FoEI was at risk of breaking up, and during which many members did not feel there was 'solidarity' in the federation.

In that year, with around sixty-eight national member groups, the network faced a crisis that challenged the very basis of the solidarity that the groups assumed they shared as members of FoEI. Accion Ecologica, the FoE group from Ecuador, resigned from the network. The event led to widespread questioning from a number of other groups about the balance of power within the network. An Extraordinary General Meeting (EGM) was called in 2003 in Colombia. At this meeting, a proposal to embark on a ten-year strategic visioning and planning process (SVPP) was approved. The central objective of this process was to explore and combine the principles and objectives of the members into something coherent, that all members could identify with.[8] Currently, seven years into the SVPP, the policy of decentralised decision-making is still considered a defining characteristic of FoEI.

In 1992, the network adopted the idea of 'environmental space' that combines the notions of environmentalism and equity (Sustainable Societies document, Amsterdam archives). This was the first occasion when the radically different approaches to environmentalism among FoE members became apparent, or rather the first time their differences caused serious concern. Members from 'Southern' organisations were very sceptical about including 'corporate responsibility' into the agenda of the network. This was exacerbated by an incident that primarily affected FoE Ecuador. FoE Netherlands, whose publication 'Sustainable Netherlands' was the model upon which 'Sustainable Societies' was developed, organised a conference to which they invited a Trans-National Corporation (TNC) to participate, in order to disseminate the notion of corporate responsibility. This TNC later used this particular event against FoE campaigners in Ecuador, urging the activists there to be 'more reasonable like the European FoE groups' (SVPP working document 2006, p. 15).

Two further factors that exacerbated this incident were a lack of communication and a lack of explicit understanding of differences in the network. It eventually led to FoE Ecuador leaving the network. The FoEI Executive found widespread support to call an EGM, held in Cartagena, Colombia, in 2003, entirely dedicated to resolving issues of communication and conflict resolution within the network. This led to the creation of the SVPP.

Before this crisis, General Meetings focused principally on campaigns, on the grounds that discussions about understandings, beliefs and ideology would impede the force of the network's campaigns (*ibid*: 15–16). However, after the various incidents with South American FoE groups, especially FoE Ecuador's resignation, it became clear that the network would be more effective in achieving its aims if it had a consolidated approach. The SVPP requires that all the FoE member groups explore their assumptions, ideas, beliefs and understandings of environmentalism and what they want the FoE network to achieve. Most FoE groups, or at least a number of activists from them, have now been participating for at least five years in a process of exploring their particular formulation of environmentalism, in order to present them at General Meetings. At these meetings, through workshops and other activities, these different approaches were combined to establish a common 'vision and mission' and a ten-year strategy. The process is planned to be ongoing and to include new organisations' ideologies or changing ideologies.

Ontological Politics in a Common World

FoE activists, especially in the context of the International Federation, work for equitable decision-making processes that are sometimes, but not always, thought about in terms of democracy. In Chapter 6, I will explore FoEI decision-making processes and structures in depth, but for the moment, it is important to unpack the historical and political assumptions often ascribed to these international decision-making processes. In terms of the ontological politics I discussed previously, collective decision-making processes are pivotal places where alterity can be subsumed in discourses of 'diversity' (Dirlik 1997: 72). These international decision-making or policy-making processes are the most obvious places where real disagreements and contestations are elided in favour of 'a gloss of harmony' (Müller 2013). In the ethnographic studies of various UN organisations in Müller's edited volume, there are many factors that also apply to FoEI practices, and that deserve the critical edge of ethnographic study. As with United Nations position papers, documents, agreements and so on (Müller *ibid*), FoEI documents are also negotiated to the letter. Through this process of negotiation, heterogeneous cultural and political realities are transformed into registers of knowledge and official documents, drafts are 'tamed' and the documents become 'polite, cleansed of their conflictive elements and rendered "technical"' (*ibid*). Although UN bodies have gained increasing legitimacy, their role in world politics is problematic. These bodies have no legislative power, and their funding has been radically cut in recent years. However, in attempting to combine the conflicting roles of standard-bearer (holding up ideals), technical overseer (order, control) and non-politicised or diplomatic (negotiating the conflicting political and economic interests played out within UN bodies), UN officials have increasingly emphasised the 'technical' register. Such UN processes and documents do not so much highlight political responsibility for current problems as offer future-oriented 'guidelines' about how best to proceed (Larsen 2013). Although the UN's agreements and position papers, as with policy documents (Shore, Wright and Pero 2011), do not have the weight of official law, as 'soft law' they create normative standards that have very real 'hard' effects when mobilised as part of projects in local contexts (Müller 2013: 10).

Unlike UN organisations, FoEI activists are not at all concerned with generating documents that are 'technical' as opposed to 'political'. However, the importance of consensually agreed-upon documents does carry similar concerns that they should be 'harmonious' and offer a unified public face, if for no other reason than to garner funds from donor agencies. Internal conflict has often been used against FoEI as a reason to deny its legitimacy. Attention to the rhetoric of FoEI's decision-making processes is essential even if FoEI aims for consistency for reasons that are different from why governments and transnational institutions attempt to portray themselves as homogeneous and internally consistent. Importantly for the question of ontological politics, Mosse (2005) argues that processes and practices such as these generate a 'transnational epistemic community' that ends up reproducing understandings that 'fit' in documents, and therefore nullifies alterity. In the context of the ontological politics so important to FoEI activists, this raises the question of whether the discussions, debates and decision-making and policy-making processes, considered by FoEI as an 'inclusive global process', in fact add up to epistemic colonialism.

In his ethnography of Yshiro narration of the global, Blaser (2010) reminds us that even in what might be considered very local contexts, the global is active in the way cultural and moral logics are translated. In the translations between Yshiro and what he calls Cartesian moral logics, there is an asymmetrical relationship. Cartesian moral logic is more powerful and it is likely that "these translations may actually deepen the grasp of one logic in detriment of the other". The translations that favour or strengthen the *yrmo*, the Yshiro cosmos, tend to "go no farther than the Yshiro communities." (*ibid*: 125). The violence of epistemic colonialism should not be underrated. I explore FoEI activists' decision-making processes, and in Blaser's ontological sense, the translations they carry out, ethnographically in the chapters that follow; however, for the moment, it is important to outline two lines of thought that reset the way this ethnography can be interpreted.

First, when considering epistemic colonialism, specifically Viveiros de Castro's (see Viveiros de Castro et al. 2014) injunction to 'permanently decolonise thought', Candea (2010) has raised the prickly question of how to 'take seriously' one's own concepts. Viveiros de Castro explains that his suggestion to take seriously is not relativism or tolerance, but implies allowing 'ontological self-determination', or as Candea explains it "refraining from either assent or critique, in order to allow the people themselves to specify the conditions under which what they say is to be taken" (*ibid*: 2). Viveiros de Castro's proposals require that the scholar refrain from attempting to explicate the other, echoing Strathern's (1990) concerns that such scholarly analyses 'explain away' people's own explanations. However, Viveiros de Castro continues that same scholar should carry on rigorously analysing and explaining things that are "near to or inside of us", which according to his definition would amount to 'not taking them seriously'. Candea highlights the asymmetry of an approach wherein things (such as visions, beliefs, ontologies) that are "ours" can and should be challenged, whereas "theirs" (meaning everyone else's) should be "taken seriously", which in Viveiros de Castro's redefinition means leaving them in a state of possibility (*ibid*: 148) rather than 'explaining them away'. Working in Corsica and with scientists in the UK, Candea is wary of proposals that in some way recast the tenacious West-vs-Rest divisions, having learned from fieldwork that what is cast as 'us', as 'close', as 'known', always offers up further difference. The 'West' is as full of difference as everywhere else. Graeber (2015: 21) takes this further. He argues that were we to consider the concepts currently associated with 'the West' we would not only find multiple conversations and interpretations (rather than internally homogenous definitions, which global categories such as 'the West' imply), but we would also find that these concepts have originated in very disparate places and periods around the world and are actually used and practiced by millions outside of the 'geographical' West. A simple example is the way that principles of individual human rights, closely associated with the enlightenment, have been appropriated by women in Bangladesh to protect themselves from violence (Ram 2006). And vice versa, the historian Steffi Metze (nd) shows how central thinkers of the enlightenment such as the Edinburgh *literati* were in constant correspondence with relatives in India in the 1700s. She suggests that the sustained engagement between scholars in Scotland and in India during this period created the very seedbed for concepts that grew from the enlightenment.

In excavating the history of the notion of democracy, Graeber 2007 shows how there never was a 'West'. In fact the tendency to oppose the ways of life of 'others' to a homogeneous 'us' has political ramifications internal to anthropology as a discipline (Chua and Mathur 2017).

Candea (2010) admits that Viveiros de Castro's proposals do not necessarily imply a homogeneous 'us', since what is required is a continuous interrogation of the conditions of possibility of a "we". Moreover, for Candea, Viveiros de Castro's proposals offer a gift to anthropology at home, or what he calls endo-anthropology. This gift is the reminder that what makes our fieldwork anthropological is the "commitment to taking seriously the multiplicities internal to what we thought was simply "us," instead of either taking these worlds for granted or subjecting them to the usual critical unveiling" (*ibid*: 150). In the context of FoEI, this is important to keep in mind since in their negotiations of what is to become of FoEI, activists face a very real need to recognise when their own positions need to be pushed forward or when they need to acquiesce, to 'compromise'. In the 'real world' of politics, taking seriously in Viveiros de Castro's terms means risking one's own position in the world. Of course, he is not arguing that a permanent decolonization of anthropological thought is the way to achieve such decolonization for everyone. However, I am committed to developing theories that are not just methodological tools for us but of value to the people we work with. This commitment means that we can't have one understanding of alterity for anthropologists and another one for environmentalists.

Anthropology and Ethnography

This book is based on spending time with people who in some way or another feel they belong to, work with or are part of some grouping either called Friends of the Earth or Friends of the Earth International. I worked with three nodes within FoEI to trace the particular experiences of these three groupings in the constitution of the international milieus of the network. The three groups are FoE Malta, FoE Brazil and the permanent international Secretariat of FoEI. At times, the people I worked with called themselves activists. In Chapter 5, I unpack the notions and enactments of personhood that make it possible for people to consider themselves 'activists'. Others do not think of themselves in terms of activism, but as employees or volunteers, and sometimes as members. Some of these people can be thought of as at the core of their respective groupings—whether national member groups or the International Secretariat. However, in the course of fieldwork I also worked with people on the fringe of such groupings who did not consider themselves, nor were considered, to be 'core'. Yet their very status and presence is central to understanding the dynamics of belonging in such assemblages. For brevity I refer to all the people I worked with as 'activists', or 'FoEI activists' and I specify when I am referring to people who are core, or who specifically do not refer to themselves as activists.

Each ethnographic chapter traces a stage that a newcomer to Friends of the Earth might follow, throwing light on the processes experienced by the activists and by which FoEI and other FoE groupings are constituted and maintained as real entities:

> Chapter 3: Asks where FoE activists are, revealing the different senses of emplacement and scale that are involved in being an activist.
>
> Chapter 4: Describes being introduced as a newcomer to a FoE national member group or FoEI, drawing on historical documents that depict these FoE groupings as supra-personal entities.
>
> Chapter 5: Shows how one's life as an activist and ongoing life practices become part of the organisation.
>
> Chapter 6: Charts developing relationships at international meetings with FoE activists from other countries who, rather than perceiving each other categorically as national representatives, get to know each other in person and thereby forge a sense of belonging to the entity 'FoEI'.
>
> Chapter 7: Describes returning to one's home country after the international meetings, and going on to develop relationships and FoEI using communication technologies.
>
> Chapter 8: Realising that, as a FoE activist, ways of relating to other institutions or entities such as the EU or other groups within FoEI depend on the particular vectors in play.
>
> Chapter 9: Explores how mission statements, policy documents and ideological terms like flatness, decentralisation, inclusiveness and facilitation are imperfect experimental tools with which to work towards the future that FoE activists hope for.

The book is organised along these lines for an ethnographic purpose. Specifically, the successive steps of increasing familiarity with different FoE groups and ways of working, in the experience of most of the activists I worked with, and in my own experience as a FoE activist, best bring to light the effects of opacity on agency and the effectiveness of institutions. However, Chapter 2 first sets out the theoretical framework for the thesis. I propose a synthesis of the approaches of Ingold, Latour and Haraway that I call ecological phenomenology. This synthesis provides a detailed theoretical grounding upon which a range of scales (micro- to macro-) can be taken into account, as well as providing a possible explanation for diverse human ways of life within a single common world. I introduce three concepts to map and explore the simultaneous workings of impersonal structure and personal agency. These are *fields of forces*, *vectors* and *direction of attention*. I propose to substitute these notions for the more traditional notion of 'relations' in anthropology. I shall propose that the interplay of vectors more precisely explains how the various types of entities (including non-humans and specifically supra-personal institutions) that FoE activists encounter are formed, the agency they exert, as well as the effectiveness of activists' personal power in dealing with them.

Although in its theoretical ambitions, Chapter 2 is apparently of a different character from that of the succeeding chapters, and is also ethnographic, albeit obliquely so. The sorts of existential questions I explore in Chapter 2, and the provisional conclusions I offer in the form of ecological phenomenology and the three notions of fields of forces, vectors and direction of attention, are characteristic of the sorts of questions activists ask themselves in the process of making their life path and forming their activism. A number of the issues and questions I raise here are the questions that led me to work and then to study with FoE, and provide therefore a reflexive ethnographic description of the questioning that drove me, and others similarly although not necessarily with the same questions or taking the same paths, to become activists. On the other hand, I also choose to position this discussion at the beginning of the book, rather than as part of the conclusions or only as the issues emerge in subsequent chapters, for a specific reason. Following the reflexive turn in anthropology, I present my theoretical framework at the very start to provide the reader with an explicit position against which they can evaluate the observations and analyses that follow.

Finally, a note on the title of this book: For the last ten years Tim Ingold (2008, 2013, 2013) has insisted on the difference between anthropology and ethnography. In Ingold's analysis, ethnography is a documentary practice that depends on a process of data collection for later analysis. Anthropology, on the other hand, is a transformative practice, concerned with the emergent possibilities of life.

Ethnography as data collection embodies certain assumptions. The first is that knowledge needs to be retrospective (Miyazaki 2004)—that the work of the ethnographer is to analyse bits of 'information' once they have been extracted from the flows of life in which they have emerged. The second is that the lives in which this 'information' arose can be explained through the ethnographer's work of putting such knowledge into its 'proper' social and historical context. This critique is very similar to Strathern's 'explaining away', and Ingold emphasises the circularity of this form of analysis. Third, and closely related to this, ethnography as data collection also privileges regimes of expertise, where only the enlightened scholar can 'see' the workings of the social, making people's own self-analysis redundant. Marcus (2009) also defines ethnographical fieldwork as being 'documentary', but his critique is aimed at the implicit instrumentality of fieldwork that is not collaborative. Combining Marcus's view of non-collaborative fieldwork as instrumental, or in Rodriguez's (2015) terms 'extractive', with Ingold's arguments about ethnography we can see how the praxis of ethnography *as* data collection is much more tenacious than most anthropologists dare to acknowledge.

Projects that are studies *of* things more often than not tend to be documentary, or in Ingold's terms 'ethnographic'. Projects that are anthropological, on the other hand, continue to develop dialogical relationships with the multiple people, places and things through which knowledge emerges. They are studies *with* things. In the light of this and my obvious interest in

Ingoldian thought, it may come across as odd that the title of the book is 'an ethnography of global environmentalism'. There are two truths here. The first is that for the publisher 'ethnographies' are more recognisable to their target readers, and the second is that in working with FoEI activists I have come to value the documentary purpose of ethnography. Indeed, Ingold explains that his intentions in distinguishing between ethnography and anthropology have often been misread. He writes,

> what I want to do is to *protect* ethnography as something of value in itself, as against its reduction to 'ethnographic method', in which empirical inquiry becomes nothing more than a means to the end of anthropological generalisation. It is this reduction that leads to the (false) identification of ethnography with 'data collection'.
>
> (pers comm)

In fact, ethnographies are currently essential 'tools' that support the anthropological stance that one's own taken for granted understandings of reality are contingent and that the world may be otherwise. As we have seen previously, institutions bent on global governance focus on 'forward looking' discourses such as the practice of generating 'guidelines' (Larsen 2013). This orientation towards the future permits international organisations to elide varying responsibility not only for past actions but for ongoing actions as well. Therefore, documentary practices are important in order to provide accounts, and therefore accountability, and critique.

In order to maintain the politically valuable aspect of ethnography, of account and critique, but minimise the problematic aspects of regimes of expertise and the embedded assumption of retrospective knowledge, I put forward this book as my response to FoEI activists' own questions. The book is an attempt to enter into dialogue with issues that both anthropologists and FoEI activists share. It is increasingly common that fieldwork subjects analyse and address similar questions conventionally asked by anthropologists (Marcus and Fischer 1999; Holmes and Marcus 2008). The result is not only that anthropologists ask slightly different questions, if at all, from the questions asked by their fieldwork subjects, but that the production of knowledge is common to both. (Ingold 2007; Gatt 2009, 2010). In fact, the most valuable realisation is that we are not primarily 'producing knowledge' in common, but working out our lives together: "We do not live in order to produce knowledge: we know in order to produce life!" (Ingold Pers comm).

Though the act of writing a description may be retrospective, texts themselves are not necessarily static. Texts are a central resource for ongoing exploration, for both anthropologists and FoEI activists. The policy documents that FoEI produces through complex processes, over many years, are alive with the conversations that generated them. The descriptions, analyses and arguments about these policy-making processes presented in this book are a means to carry forward a dialogue, with other scholars, students and also with FoE activists. This is a broad and multi-stranded dialogue, not only involving academic texts being read 'outside' universities but also activist research and work informing research interests and scientific directions. FoE activists have, in effect, been engaging in experiments in ontological politics for decades. Academics have highlighted how activists have had constitutive roles in shaping not only transnational political debates (Boissevain and Gatt 2011) but also participated in shaping academic interests, for example, genetically modified organism (GMO) sciences (Lister 2003). In many cases, the distinction between activism and academia makes no sense at all.

Recent ethnographies of NGOs emphasise the ongoing negotiations of NGO members, often with the explicit aim of deconstructing the rigidity that organisational structure has been understood to impose on human NGO actors.[9] This ethnographic scholarship responds to the body of social theory that takes structure and agency as its focal point, specifically in the writings of Bourdieu and Giddens. While this literature has given due attention to the creativity of human actors, I argue that it is also necessary to pay attention to 'structure' and to the supra-personal actors that are perceived to emerge from structure. In other words, I aim to take seriously FoEI activists' experience and understandings of structural power by proposing a trio of concepts: fields of forces, vectors and direction of attention, which I introduce in Chapter 2. The questions that academics are repeatedly faced with are: How are their analyses relevant? What is the impact of their work? A fundamental insight emerging from feminism and post-colonial studies is that any action, choice or piece of text is inherently political and situated. Academic analyses and academic texts *are* political; they *are* activism.

Notes

1 www.foei.org/en/who-we-are accessed 6th April 2017.
2 Except in the UK, where there is a FoE Scotland and a FoE EWNI (England Wales Northern Ireland) and in Belgium, where there is a FoE Flanders and a FoE Wallonie.
3 http://news.nationalpost.com/news/canada/save-most-of-the-whales-greenpeace-now-supports-inuit-hunting-but-native-groups-still-wary accessed 6th April 2017
4 See Gatt 2017 on the multiple implications of voice in relation to ontological politics.
5 While there are a handful of anthropological studies of the ideology or logic of environmentalism (Milton 1993; Argyrou 2005; Ingold 1996), most anthropological research into environmentalism up to now has focused on subordinate groups, or on the inhabitants of a particular delimited area who are faced with environmental issues. Other studies explore localised environmental NGOs (Berglund 1998; Choy 2005; Gatt 2001), or the rhetoric and practice of environmentalism dispersed across different NGOs or looser groups, and non-activists (Milton 2002; Fortun 2001; Tsing 2005). There are also a number of studies of the politics and practices of NGOs, such as Riles's (2000) study of Fijian women bureaucrats and activists, Hilhorst's 2003 ethnography of an NGO in the Philippines, as well as

the contributors to Bernal and Grewal's (2014) volume *Theorizing NGOs*. These volumes inform this work, especially with their attention to the everyday politics and creativity of the activists we encounter in these ethnographies. However, none of them focuses on environmental activism, and I believe it is the question of the environment that makes the work, struggles and thinking of FoEI activists fertile ground through which to think about current global crises.

'Studying-up' is a growing area for research in anthropology and the result is that government institutions have received increasing attention in recent ethnographic work (Abélès 2004, Deeb and Marcus 2011, and the contributors to Müller's 2013 edited volume on multilateral transnational institutions). However, the same cannot be said of donor agencies (Anderson and Berglund 2003: 11) or of international NGOs, barring Hopgood's moving ethnography of the international office of Amnesty International and Graeber's (2013) work with the Occupy movement. Little attention has been directed towards international environmental NGOs (ENGOs). FoEI is one such organisation. More specifically, as I described earlier, it is an international federation of 'grassroots' ENGOs. FoEI is one of the prominent ENGOs in international as well as in some national contexts.

6   The approach of multi-sited fieldwork purports to make globalisation accessible on the scale of human encounter (Strathern 1996; Marcus 1995; Falzon 2009). There are studies of transnational work such as Hannerz's (1996) ethnography of journalists, Garsten's (1994) study of Apple Inc., and Wulff's (2000) exploration of the transnational world of ballet. Numerous studies of migration and diasporas have contributed to how we think about lifeways that stretch across different geographical places (Rouse 2002), and how to focus on the relations between places (Falzon 2004). Kim Fortun (2001) has studied the Bhopal disaster of 1984 which, she argues, was a happening that extends and spills over simple territorial boundaries, making it necessary for her to follow Bhopal to the Unites States, to other cities and over diverse time scales.

7   Paris Agreement 2015 http://unfccc.int/paris_agreement/items/9485.php accessed on the 10th April 2017

8   The source of the information about events that led to the commencement of the SVPP is from an SVPP working document, descriptions of the happenings by Marijke Torfs at the FoEE AGM 2006 and Doherty (2006).

9   Questions of structure and/or, agency arise frequently in ethnographies of NGOs. This is unsurprising since the organisational structure of an NGO is considered a distinguishing feature of this form of social association (Hilhorst 2003). NGOs function in what Lister (2003: 178) calls 'legitimating environments', the social and institutional milieus that lend legitimacy to NGOs. Within so-called Northern legitimating environments, NGOs are also delineated according to their organisational structure (ibid.). Social associations that are not organised in a formal sense are rarely recognised as part of civil society (Hann 1996: 13;). Therefore, the structure of associations plays a part in creating the content of the actions, relationships and understandings of NGO actors. 'For NGO actors, the legitimation of their organization is a matter of (organizational) survival' (Hilhorst 2003: 8). In many organisational ethnographies of NGOs, however, their authors have focused on the negotiations of the day-to-day lives of NGO actors, or in other words on their agency (see for instance Riles 2000; Hilhorst 2003; Hopgood 2006; Mosse and Lewis 2006; Yarrow 2011), rather than on the constitution of such organisational structures.

2 Proposing an Imaginary
Fields of Forces, Vectors and Direction of Attention

FoEI was set up to mirror and improve upon supra-national decision-making bodies such as the UN. Many FoE groups have shifted away from such bodies and now focus on 'support for communities'. Accordingly, the representations of FoEI groups have also come to include 'support' and 'creation' alongside the original labels of 'opposition' and 'stewardship'. The recently coined motto says it all: 'Mobilise! Resist! Transform!' The implication is that activists within FoEI are understood to be capable of different forms of power, such as resistance as well as generative power. What power is, and how different types of power have bearing on each other, is the focus of this chapter.

When working with FoE groups, it was impossible to ignore that questions of power are not restricted to human affairs. The negotiation for and *with* non-human aspects of the environment underpinned the ideas, the campaigns and discourses of FoE activists. I talk about negotiation along the lines proposed by Latour (2003) who argues that non-human things communicate by the way they respond to their surroundings and the occurrences therein. Latour (*ibid*) refers to this imperfect, and yet functional, way of communicating as 'speech impedimenta'. Although FoE activists do not talk of speech impedimenta or of negotiations with non-humans, their practices and ideologies resonate with the notion that non-humans are forces to be reckoned with, not passive resources dependent on human cognition and action to give them existential power. Sometimes these non-human forces are powerful and need actions that mitigate their effects, such as in issues of climate change. At other times, these non-human aspects are vulnerable and need protection from human actions, especially in issues such as loss of ecosystems due to urban sprawl or industrial silviculture. In the deep green politics of FoE Malta, the Earth is to be protected for its own sake (and not only for what it can provide for human needs). ProNatura/FoE Switzerland initiates lawsuits in the name of particular ecosystems. FoE Colombia struggles to protect communities' rights and knowledge of their ancestral land from being sold off to mining companies through property laws that only recognise the individual. NAT/FoE Brazil runs 'extractive reserve' projects on the edges of the Amazon forest where the indigenous populations work together with environmentalists to counter deforestation without forcing the inhabitants to leave the area; extractive reserves are not 'Nature reserves'. Every FoE group in one way or another includes both humans and non-humans as an active part of political negotiation, and they explicitly fight against regimes that regard both human beings and non-humans as passive resources to be exploited. All the activists I met, even those for whom English is their mother tongue, talk of mineral 'exploitation', rather than mineral 'extraction', in the same way that they talk of human exploitation.

Planting eucalyptus in the Pampa is problematic partly because as fast-growing trees they consume a lot of water. The tree is granted its agency. Another reason why eucalyptus plantations, for paper pulp, are problematic—for instance, for FoE Brazil and the grassroots communities they represent—is that the plantations radically change the landscape. 'Gauchos are used to wide open views of the Pampas. What will having plantations of one of the tallest trees in the world do to us?' one Brazilian activist asked at a public hearing. Traditions and memory, most often considered part of the 'social' domain, in the sense that they are cut off from material issues, are understood by FoE activists never to be separate from the material world. The traditions and memories of Gauchos grow with the landscape they work in and with (see Okely 2001). Being part of the environmental justice movement, for FoE activists, humans are always understood as part of the world. Nature is not a human-less domain opposed to culture. The powerful forces of climate change need to be negotiated with for the sake of people as well as the non-human parts of the ecosystems they live in. The Pampas need to be protected from mass eucalyptus plantation to protect the traditional livelihoods of the countryside Gauchos as well as to protect the Pampas' 'rich and unique' flora and fauna.

In the workings and understandings of FoE activists, non-human aspects of the environment are enlisted in people's projects. Sometimes they cooperate, at other times they are recalcitrant (Latour 2003, 2005). And yet in established, and influential, social and political theory, non-human things are typically portrayed as passive, as resources (Giddens 2009; Bailey 1969). 'Allocative' resources (Giddens 2009) are used by people to construct their social worlds, where 'social' here implicitly denotes a supra-natural domain. As Pfaffenberger points out, the agency of non-human aspects of the environment is 'all but invisible through the lenses provided by Western economic, political and social theory' (1992: 500).

However, recently non-human elements of the world have become more central to some streams of scholarship. Actor Network Theory (ANT), for instance, has championed the notion that rather than being entirely the preserve of humans, agency and power inhere in 'heterogenous associations' of humans and non-humans (Latour 2005; Knappett and Malafouris 2008). Power is located in the network, not in finite individual entities, such as humans, states or societies. Here, a network is not to be understood in terms of the Social Network Theory of the 1960s, in which the nodes of the network included only humans (Riles 2000) and agency was not conceptualised as 'dispersed' in the same way (Latour 2005). In ANT what is meant by 'humans' is also questioned; humans are also made up of such networks (Law 1992). ANT understands agency to result when the interests of various actors in a network are aligned. Agency is thus distributed. Individuals or nodes in the network only act. It is how those actions align with others that affect the outcome. The term 'actant' denotes the simultaneous condition of being active and being acted upon, which is the condition that arises when interests are aligned. In other words, when the actions of actants cooperate, the network has effectiveness (Law 1992).

ANT has several pitfalls. Since agency is dispersed in the network, ANT seems unable to account for 'intentionality' and asymmetries of power. Conversely, with the model of alignment, ANT can neither explain 'unintended consequences of action', where powerful structures are constructed with no necessary relation to 'interests' (Giddens 2009), nor other possible configurations of interest, resistance, indifference and so on. On the one hand, ANT creates a robust framework for the participation of non-humans, so important to the politics of FoEI. On the other hand, however, by not addressing intentionality, it does not offer a satisfactory framework to the classic problem of structure versus agency, with which FoEI activists, in their struggle to 'make a difference' in relation to multinational corporations, national governments and international bodies, are in practice all too familiar.

In order to understand how FoEI is constituted and maintained, a theoretical framework is needed that allows for both impersonal structures and personal power. Because of my studies, I am proposing 'fields of forces' as this framework. In order for this frame to accommodate intentionality, agency, structure and unintended consequences, I have developed two concomitant notions: vectors and direction of attention. These notions are themselves based on a particular ontology, which derives from a synthesis of the works of three scholars, each of whom contributes particular understandings and ways of thinking. In the section that follows, I first describe the principal notions that I draw from Ingold, Latour and Haraway and how they resonate. I call this synthesis 'ecological phenomenology', extending a term used by Eriksen (2006). The framework of force fields, vectors and direction of attention, as well as the ecological phenomenology they are based on, are necessary to make sense of the way that non-human and human phenomena are intermingled in this book. This mixing depends on a particular definition of 'sociality'. In ecological phenomenology, sociality does not refer to a domain of human interaction that is disconnected (or that constructs) the material domain, as it is in Durkheim's sociology. Rather, Ingold describes this approach to sociality thus:

> [I]t is opposed to an ontology of the particulate that imagines a world of individual entities and events, each of which is linked through an external contact—whether of spatial contiguity or temporal succession—that

leaves its basic nature unaffected. In the terms of the physicist David Bohm (1980), the order of such an imagined world would be *explicate*. The order of the social world, by contrast, is *implicate*. That is to say, any particular phenomenon on which we may choose to focus our attention enfolds within its constitution the totality of relations of which, in their unfolding, it is the momentary outcome.

Ingold (2007c: 18)

Of course, the implication of this could be that to understand any phenomenon we would need to take into account the entire universe. On the one hand, environmentalist ideology does not exclude that (Argyrou 2005; see also Chapter 3). In fact, I brought that ideology with me to this study from previous work with FoE Malta, and subsequently found similar notions in the work of Ingold and Latour. On the other hand, the ever-expanding definition of what is social creates challenges for fieldwork. If everything is generative and constitutive, how can one find a focus for research, other than a list of the elements (at least those that are perceivable, anyway)? I deal with this question in more detail elsewhere (Gatt 2010). However, the working proposal that made it possible to sit down and write anything at all was the recognition that happenings draw our attention to particular aspects and players, in a process that I have called 'serial closure' (*ibid*). Serial closure is not so different from the idea that long-term participant observation allows researchers to follow the flow of happenings around them (Malinowski 1922; Coleman and Collins 2006). What is different is that the ontology and frame of fields of forces, which I describe in detail next, permits a consistent imaginary of *how* the different players, in the phenomena we have our attention drawn towards, have real and transformative effects on each other. An explanation that other theoretical paradigms, such as structural-functionalism, structuralism, and social constructivism fail to do satisfactorily.[1]

In what follows, I do not argue that together Latour, Ingold and Haraway offer a coherent theoretical paradigm. Rather the resonances between them offer a synergy that is helpful in trying to understand the constitutive forces of particular situations, including divergent perceptions of the same environment, without the trade-off of having to leave politics, ethics and an engaged approach out of the analysis. The second section presents in detail the notions of force fields, vectors and direction of attention.

The third section is dedicated to a critique of a recent work on power and politics. I explore in depth the work of Nigel Rapport (2003), who revisits the question of existential power and offers important political and ethical keys to the question of intentionality and personal agency. His framework, however, does not allow for the agency of non-humans. I dedicate considerable space to this close reading to highlight the underlying assumptions of power and politics that prevent a seamless attention to ecological questions. And this is necessary, because paying insufficient attention to the multiple constitutive relationships through which life is possible precludes ethno-graphic understanding of *how* some actions have more or different effects than others.

Although, Rapport may be considered unrepresentative of much current anthropology, few anthropologists elaborate in detail the underlying ontol-ogy and epistemology that informs their work. As a result, the material or the non-human are often 'added' to anthropological accounts, which may lead to contradictions and ethical issues in more mainstream works such as the ones I analyse further on in this chapter.

A note on 'metaphor' is required at this point. Fields of forces, vectors and direction of attention are conceived in this book to be both metaphori-cal and metonymical. They are metaphorical in the sense that the imaginary evoked by a vector is much simpler than the processes and things I will use it to describe. In using force fields to portray different factors at work in peo-ple's daily lives, I am abstracting from unimaginably rich details to the lines of vector diagrams. I use these notions to conjure up imaginary landscapes by means of disparate domains—that is metaphorically, that are simpler to convey, and leave it up to the reader to fill in the details. Together these concepts—force fields, vectors and direction of attention—are an attempt to create an imaginary in which intentionality, unintended consequences and non-human actors can be coherently taken into account. However, I intend the *qualities* that force fields allow me to describe to be understood as more than simply metaphorical. In my account, force fields, vectors and direction of attention, although presenting what goes on in the world in a simpli-fied form, are nevertheless *part* of the world. In this way, these notions are metonymical: force fields are part of how the 'real' world works that act as substitutes for the purpose of (partial) description.

Ecological Phenomenology

In parallel to ANT, Tim Ingold has developed an ecological approach that sometimes resonates with ANT, in that his approach also does not privilege the human to the exclusion of non-humans. However, Ingold's approach is radically different from ANT when it comes to the under-standing of relations (Ingold 2008b). In ANT the explanatory metaphor is the network. Human individuals and non-humans are nodes in the net-work. However, it is not clear whether the lines that link the nodes in the network have actual presence in the world, and if they do what that presence may be (*ibid*). In addition, the widespread notion of hybridity in ANT indicates that these nodes in the network are understood as ready-made, unchanged by their participation in the network. A hybrid is a thing combining at least two elements. It is necessary to imagine those elements as bounded in the first place in order to have a 'hybrid'. This contradicts the premise in that being an actant assumes being acted upon as well as acting (Ingold 2009).

Ingold (2000, 2007a, 2007c, 2008a, 2011, 2013, 2015), on the other hand, understands all things to be in a process of mutual constitution in a field of unfolding relations. There can be no hybrids, understood as the combination of two fixed things, because things are never fixed, never completed. This is where Latour resonates more with Ingold than with other ANT writers such as Law (1992) and Mol and Law (1994). Latour's *Politics of Nature (2003)* offers an insight into how to handle the question of intentionality and the ethical problems of dispersed agency. Donna Haraway (1991, 2003) adds a further dimension to the question of politics and power, making space from the start for semantics, symbolism and meaning, as well as attention to the material world. She also counters ANT's focus on aligned interests with her militant attachment to 'dirt', to inharmonious voices and positions (*ibid*).

The works of all three—Ingold, Latour and Haraway—converge on at least four points and these points together form ecological phenomenology. First, all three claim that human beings perceive the world directly. This is in contrast to the broadly constructivist view[2] of perception according to which humans receive inchoate stimuli from the outside world through the senses, which their minds then organise according to the patterns of a learned culture. For all three theorists, the world is not a blank canvas upon which we inscribe our conceptual constructions (Ingold 1992, 1993; Latour 2003; Haraway 1991).[3] Ingold (1993) shows that without direct perception, no learning would be possible at all. Moreover, things often resist our conceptualisations (Ratto 2009). Perception and knowledge can only ever be partial and situated because they are constituted relationally (Ingold 1992, 1993; Haraway 1991; Latour 2003, 2005). In other words, perception is not a process of reception, where our senses receive stimuli from a ready-made world. Rather, things emerge from manifold relations in an environment, and once they emerge, they are instantaneously involved in the ongoing coming-into-being of the world.

Second, all things are constructed relationally as the outcomes co-constitutive relations among heterogeneous constituents (Ingold 1993, 2000a; Latour 2003, 2005; Haraway 1991, 2003). Latour has sometimes been misunderstood as supporting the notion of multinaturalism (Viveiros de Castro 1998)—which takes seriously the ontology of the people De Castro worked with, reproduces, in its advocacy of plural natures, a relativism that is unproductive for establishing due process for politics (Latour 2003: 29). To overcome this misunderstanding, Latour advances the idea of a single arena, oriented towards engagement, in which the single collective is continually being made (*ibid*: 30).

Third, because of this co-constitution it is important that research accounts for diverse relations, not only among persons (Haraway 1991: 198, Latour 2005, 2003; Ingold 2000b). In fact all three theorists comment specifically on the false division between human history and natural history (Latour 2003: 123; Ingold 1999: 252; Haraway 2003). Limiting the scope of social relations, in Durkheimian fashion, to a domain of human society abstracted from other aspects of the world, only leads to the tautology that the social is powerful since it is social in the first place (Latour 2005: 65). Fourth, all relations are processual; Ingold, Haraway and Latour put forward an ontology of dynamic becoming (Latour 2003; Ingold 2000a, 2007a; Haraway 1991, 2003). For Latour (2003) the world is a process, a progressive composition of the collective as it is, in Ingold's (2006: 9) words, 'a world-in-formation'.

There are also clear divergences in their work. For instance, Haraway's (2003) use of the notions of nature-cultures and hybrids, in which earlier nature:culture, human:nonhuman divisions linger, does not sit easily with the idea, in both Ingold and Latour, that everything is the result of mutual constitution (see Ingold 2008, specifically against the notion of hybrids). On the other hand, Latour does not address Ingold's (2006) critique of the notion of agency, namely that it attributes vitality to objects that are assumed to be ready-made and rendered lifeless through their excision from the flows of materials in making. In the *Politics of Nature*, Latour qualifies his notion of 'things' at length. In fact, both Latour (2003: 54) and Ingold (2000) define 'things' not as ready-made objects but as sets of relationships. Both Latour and Ingold refer to the etymology of the word 'thing' lying in public process and discussion. In a footnote Latour (2005: 65) also refers to James Gibson's notion of affordances to describe the relational quality of things; affordances being a notion that Tim Ingold (1992) introduced into anthropology. However, the notion of objects, unqualified as 'things', in *Reassembling the Social* (2005: Third Source of Uncertainty: Even objects have agency'), coupled with his use of the term agency, risks perpetuating the very ontology that splits the world into domains of mind and nature, human and nonhuman, subject and object which, in *Politics of Nature*, he argues against. Haraway's (1991: 190) argument for the need to see from a number of different situated knowledges offers a practical approach to exploring questions of power and inequality, which Ingold, with his ecological focus (Ingold 2000, see also Lee and Ingold 2006) has not set out explicitly to address, but which are not in contradiction with an ecological approach (Ingold 2006).

Their different foci are what give strength to the synergy; they do not replicate each other's work, which is what makes them complementary. Ingold's work elaborates an ontology that explains the processual and mutual constitution of our world (including ourselves). Whereas Durkheimian accounts are largely tautological because they cannot explain the emergence of power and social relations, Latour (2003) has elaborated an understanding of the institutionalisation of essences, however transient, that allows for stability as well as change. Finally, Haraway's (1991, 2003) work offers concrete examples of how political issues as well as semiotic aspects of life can be taken into account.

To sum up, ecological phenomenology proposes an ontology. That ontology can be described thus: anything that engages in the world mutually shapes and is shaped by its manifold field of unfolding relationships (Ingold 2000: 187). Such relationships occur in a single continuous world (Ingold 1993; Latour 2003) where persons and actors are constituted by the sum of the relations in which they are immersed: this includes relations with other persons, but also with the air, the ground, non-human animate beings and inanimate persons or things, in other words the ingredients of an environment. Following in the Heraclitean tradition, this mutually formative, unitary world does not 'descend into a formless mass' because difference and distinctiveness come about due to the uniqueness of position; no two sets of relationships can simultaneously occupy exactly the same position. The different positions in the world give rise to different characteristics— themselves in constant mutation and at different rates of change—that Ingold, borrowing from Gibson, calls affordances (Ingold 1992, 2008a; Gibson 1979).

It is important to note that Ingold's use of the concept of affordances avoids the deterministic undertones of Gibson's usage. For Gibson, an object affords what it does *because of what it is*, regardless of whether the affordance is perceived by anyone or anything in the vicinity. In this sense, he is a *realist*. Ingold's approach, by contrast, is *relational* (as are the approaches of Latour and Harraway). Therefore, Ingold's affordances are not fixed 'objective' attributes. Rather, as I noted previously, perception is a mutually constitutive process, as are other processes of action.

In the process of engaging with the world, people's attention is educated *towards* different affordances in their environment (Ingold 2000: 354). I extend this understanding of education of attention by adding the idea that people's attention can also be *distracted* from other affordances. This leads me to the idea of direction of attention, which I will describe in detail next.

Force Fields, Vectors and Direction of Attention

For ecological phenomenology, the world is a continuous field of constitutive relationships and all actors participate in the ongoing formation of this common world.[4] All entities and essences, animate and inanimate, are provisional, open to revision. Matters of fact return, depending on what is taken into account, to being matters of concern, only to be reinstituted as matters of fact, and so on and so forth (Latour 2003). The characteristics and affordances of such 'things' may appear stable because only certain affordances are taken into account and are therefore instituted as essences. They do make it possible for persons to choose which affordances to take into account.

The relationships actors engage in also have a direction and magnitude: they are vectors. Relationships are vectorial; actors invest a certain amount of energy (a lesser or greater degree of force) into one relationship and away from other potential relationships (directions). Here is the direction of attention: if an actor directs their vector, their force, with a certain direction and a magnitude, then those other directions to which no force is being directed can be imagined as 'unprotected backs'. Imagine Andrea, Mariangela, two FoE Brazil activists, and myself walking through the streets of Porto Alegre, on our lunchbreak. All three of us are intent upon conversation. Our attention is directed to the other people and what they are saying. Another part of our attention, though less, is also directed towards where we are walking, crossing streets, avoiding the hanging roots of Ficus trees. We are not paying attention to the bright adverts, deliberately placed at eye-level. Marketing studies have found that these advertisements have the greatest impact when people are not focusing on them. Focus makes it easier to question them. By not directing attention to the adverts at that point in time Andrea, Mariangela and myself had unprotected backs to their subtle messages. Although in this example I refer to what is present in our peripheral vision, unprotected backs and direction of attention do not refer only to the visual modality but to all possible ways of perceiving and paying attention.

The vectorial forces of other actors fill the gaps. New gaps are always being formed by the changing directions and intensity of force of different actors. The vectorial quality determines the possibilities (which may be incalculable) for interaction, but gaps, like paths, lead to new interactions. Therefore, vectors also allow for change. In other words, actors 'recognise' the different affordances of the field, and direct their attention to one or more paths of possible interactions.[5] Actors could act in all directions, as from the epicentre of a sphere. In fact, charged particles emit the same force in all directions. However, since we are immersed in relationships even the magnetic fields around these particles tend to fluctuate, to have different vectors at play. Nonetheless, we can imagine at least five, non-exhaustive and non-exclusive types of directions, which could constitute vectors: towards, away from, past each other, alongside each other, and centripetal, or shifting field. This applies equally to molecules, tectonic plates and humans as instances of provisional actors:

1   When acting towards each other, the magnitude of force will determine the path of interaction. If actor A exerting a force of 2 directly towards actor B who is also exerting a force but only of 1, the two bodies will travel in the direction of A's force. If the force is not applied directly at actor B, slightly to one side, for instance, actor B's force of 1 may be sufficient to deflect it. One of the myriad possible alternatives is that the

Figure 2.1 Acting away from each other.

forces aimed at each other are porous, allowing each other's force to permeate and therefore interact in different parts of the actor.

2   When an actor's attention is pointed away from other directions, this leaves a gap that can be filled by other forces, 'unprotected backs'— therefore, closing off certain avenues for action (but nevertheless leaving many others open).

Figure 2.2 Acting towards each other.

3   Acting past each other is when interaction does not occur; for instance, the lack of oxidisation of gold, since gold and oxygen do not interact.

Figure 2.3 Acting past each other.

4   When acting alongside each other, the forces of actors A and B are combined, creating a greater force. This is where ANT's 'aligned interests' in a network would be found.

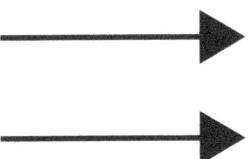

Figure 2.4 Acting alongside each other.

5   Turning in on oneself in a way that does not deny the vectors of other actors but introduces a different dimension, or topology, in which these other vectors have no or little effect. This direction builds on the idea of more dimensions than regular spherical 3-D, like nodes which can have as many directions as they have relationships (Latour 1987).

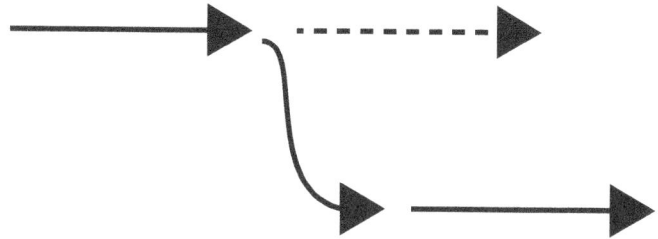

Figure 2.5 Turning in on oneself.

Ubiquitous Relations

The notion of relationship is fundamental to anthropology. Indeed, anthropology's classical focus on small-scale societies depends on the assumption that what differentiates them from complex societies are different types of relations: personal relations in the former and institutional or bureaucratic in the latter (Strathern 1996: 15–16). Anthropologists now also study in complex societies, nonetheless, as Yarrow argues, 'personal relationships are here taken to consume modern or global elements and subordinate these to their own logics' (2006: 100). In fact, with modernity and globalisation the interest in relations has become even more prevalent.

Accounts of globalisation narrate an ever-increasing volume and density of connections, and analysts of modernity detect connection where modernity presents separate categories. Here, I am referring to Latour's notion of hybrids (1993), which he claims are associations, *relations* between disparate entities (such as a microbe, what is considered disease, a laboratory and other experiments) that work together to sustain a disease like Brucellosis. The different elements are understood to be connected in a network (Latour 1993: 10).

For Strathern (1996: 522), the image of the network is appropriate to the task Latour and other ANT scholars require it to do: any description of things can be connected in a network. The possible associations between humans, non-humans and any other possible entities that can be taken into account in the logic of ANT seem ever expanding. Strathern argues that the notion of the hybrid conjoined with that of the network is 'auto-limitless' (*ibid*). She argues:

> [O]ne can always discover networks within networks; this is the fractal logic that renders any length a multiple of other lengths, or a link in a chain a chain of further links.

> Strathern (1996: 523)

The turn towards relational approaches in anthropology such as Latour's, but also including other ecological approaches such as Ingold's, has led to a proliferation of 'relations', to the extent that a recent session of the Manchester *Group Debates in Anthropological Theory* proposed the following as the statement for discussion: *The task of anthropology is to invent relations*.

Relations may be everywhere, hidden by pure forms in modernised life-worlds (Latour 1993) or the main quality of reality (Viveiros de Castro 1998; Bird-David 1994; Ingold 2000, 2007). Indeed, in non-European relational ontologies what is required of humans is to limit the possibilities of relationship (J. Weiner, quoted in Strathern 1996: 529). However, the notion of the relation is insufficiently precise. Apart from being called upon in all the diverse situations I mention above, Ingold (2007a) also points out that relationships can take at least two entirely different forms. On the one hand, there are relationships described by ANT as lines between entities forming a network. These relations are *between* already constituted entities or elements, what Ingold refers to as 'point to point connectors' (*ibid*: 75). Ingold contrasts them with relations that proceed *along* the ongoing flow of life, in which all the elements are both making and being made by those interactions (*ibid*: 103).[6] Yet both are 'relations'. I suggest that the notion of vectors is more precise than that of 'relation', because even in a network what is really going on is better described by vectors and directions of attention, in fields of forces.

I have referred above to Strathern's (1996: 253) point that the auto-limitlessness of the 'hybrid plus network' allows anything to be taken into account. A simple adding of elements does not necessarily have 'a point . . . a stopping place' (*ibid*), what she refers to as 'cutting the network'.[7] However, thinking of networks through the notion of vectors, we are driven to ask: how much force do the different nodes exert, and does this force emanate in equal measure or in the same direction along the lines (the relations) connecting them? Immediately, the notion of the vector requires one to ask for more detail about what is happening in the relationship between those elements that have been added to an account. The vectors belonging to elements in a network, not only those sustained by human effort, *punctuate* the network because they have their own ontological presence in the world. In response to Strathern, networks are not only cut by human institutions, they are also cut, punctuated or carried along by non-human forces. Vectors allow us to map and explain why some networks, or nodes in networks, have more effect than others.

The same can be said about questioning *along* relations in terms of vectors. If we consider the many paths of movement made by different persons and animals (carrying things with them), winds (carrying things in it), electricity (transducting voice and information) that weave this way and that through the world, it is not clear why some places become meeting points while others do not. However, if one considers the effects of the vectors at play in these movements, we can identify when there are collisions, avoidances or collaborations among other possibilities. Therefore, we can consider when and why those differently loaded and differently directed relationships worked together, against each other, without relationship to each other and so on. Importantly, the notion of vectors in fields of force allows us to take into account rhetoric, symbolic as well as material aspects of a situation.

Why the Need for Force Fields?

It is not enough to simply add non-humans or materials to so-called 'social' domains without unpacking the latter notion. The ontological premises of the Durkheimian understanding of the 'social' or the Schutzian lifeworld— the one to do with the collective, the other with the individual—*cuts* people's understandings, beliefs and memories away from the material, the non-human. There is an unbridgeable gap between meaning and material in the ontological premises of these logics. The gap means that when symbolic power is added to material power (see, for example, Eriksen 1998), or when ecology is added to psychological understandings of affect (see, for example, Milton 2002), without first unpacking those underlying assumptions, then contradictions are hard to avoid. In these accounts, we are left with the question of *how* such different types of power (symbolic or material, as with Eriksen 1998), or different personal states (affect and perception, as with Milton 2002), come about, or what difference does it make to recount the non-human in people's lives (Berglund 1998)? And again, *how* does change come about?

For instance, Eriksen (1998) argues that in Mauritius, people's symbolic power comes from the node in the network of Mauritian society that they occupy. In Eriksen's account, the nodes pre-exist the people who occupy them and their actions have no effect on the structure of nodes themselves. Reading his account the reader is left wondering *how* different nodes (possibly competing) appear at all. In Eriksen's account, it would seem that political changes in Mauritius are caused by 'external' factors, such as the declining demand for sugar in international markets that changes what sorts of workplaces are available, that in turn change the interaction between the different 'ethnic' communities as they are imagined in Mauritius. People and their actions seem to have no effect. And yet, Eriksen presents the locally imagined ethnic group of 'coloureds' as occupying a particular, powerful position in Mauritian society because they predominantly work in the media and arts sector. Again, we are left wondering how this came to be. Referring to power as the addition of symbolic capital and material resources does not explain how some persons and groups manage to have effect, to make things happen, whereas others do not.

A second instance is Kay Milton's *Loving Nature* (2002). Milton has applied this additive approach in trying to understand how environmentalists relate to the world. People, she argues, are organisms relating to other elements in their environment (2002: 1). This includes but does not give *undue* centrality to other people in that environment (*ibid*: 2). Milton sets out to explore how people perceive particular affordances through their engagement in the world. Mainly through the work of psychologists such as Damasio and Neisser she concludes that through engagement in the world people come to have needs and therefore anticipate finding affordances in things that satisfy these needs (*ibid*: 44). These needs are emotions that, according to Damasio's definition, are the effects of perception on the body, while feelings are the mind's conscious perception of those emotions (*ibid*: 67). Milton presents phenomenological and ecological perspectives as theoretical tools with which to understand how people perceive certain affordances in their environment and not others. However, her argument seems to return full circle upon itself. Affordances are perceived because of prior engagement that creates anticipations and directs our attention to the same affordances, reinforcing or changing previous perceptions. Furthermore, following Damasio, Milton argues that what distinguishes humans from non-humans is the conscious recognition, in feelings, of bodily perceptions or emotions. Our experiences of the world are in turn perceived and made sense of by our consciousness (*ibid*). This argument seems to reinstate the Cartesian position that humans are defined by their conscious mind, that Milton herself argues against in promoting an ecological stance. A possible reason for this apparent contradiction is that the mutually constitutive principles of the 'dwelling perspective' have been referred to but not entirely incorporated into the argument. In Milton's book, there is an insistent search for a causal relationship, determined by a single primary agent. Yet a fundamental premise of the dwelling perspective is that the conscious perception of emotions cannot be seen as a second step in a causal chain according to which the body first receives perceptions as emotions and the mind then perceives these emotions as feelings. Rather, the mutually constitutive standpoint of Ingold's ecological phenomenology would posit consciousness as one aspect of a *simultaneous* process. The radical innovation of ecological phenomenology is that the organism engages in *ongoing multifaceted formative* relationships.[8]

The idea of fields of forces is intended to provide a tool to describe the negotiations, oppositions, collaborations and avoidances that create effects in the world, that make things happen. However, it is also intended to help develop notions of power and agency that allow for both personal responsibility and effects that cannot be reduced to the actions of individuals (but maybe to the actions of many).

Inherent Contradictions

Rapport's (2003) book *I Am Dynamite* is dedicated to developing an 'as if' thesis on existential power. His book is valuable for the propositional stance it offers anthropologists; social theory participates in shaping the world, so to write about how people are determined has the effect of limiting their agency. According to Rapport's 'as if' stance, writing a book on the personal power of people should have the opposite, empowering effect. However, through a close analysis of the main premises of the book, I show the contradictions inherent in descriptions of power that do not take into account non-human aspects of the world. As a result, accounts such as Rapport's cannot explain how oppression actually does happen. In what follows the italicised phrases are paraphrased from Rapport.

The Self Is Transparent, the Other Is Opaque

According to Rapport, interaction between individuals is not 'true' communication; rather, such interaction is the coincidence of individual life projects. The trajectories taken by individual life projects may be in alignment, even in collaboration for a while (Rapport 2003: 255). However, what may seem to be interaction is merely an individual's adaptation to the predictable habits of others (*ibid*: 220). Between individuals, Rapport claims, there is nothing except forms—empty unless individuals give them meaning. This view arises from a Schutzian understanding of the 'lifeworld'. In this formulation the main characteristics of the lifeworld are: that it is the total sphere of subjective experience; that is only accessible to the individual subject, and that it is the only source of knowledge and perception. For this reason, Rapport argues that the most readily accessible, if not the only accessible, source of knowledge and perception is the self, one's own individual lifeworld; therefore, it is *transparent*. Other humans' lifeworlds, in contrast, are the most impenetrable, the most *opaque* worlds of perception; and the breach—from self to other—cannot be bridged (Rapport 2003: 218).

If other individuals' lifeworlds are so entirely opaque, where does this opaqueness begin? What is the transition from transparency/accessibility to opaqueness/inaccessibility like? Does the self have access to a sphere of perception, a sort of aura around the body within which perception can occur, where for example extensions of the body, such as a walking stick (Bateson 1973: 429), extend that sphere of the self's perception? At what point in the transition from self to non-self is perception brought to an end by opaqueness? And most importantly, what is that opaqueness? Is it a void? If it is not a void, why is it opaque to our perception?

For the self to have any perception at all there needs to be continuity between the perceiving body and the world; without the medium of air, we could not breathe, let alone see or hear (Ingold 2006: 7, 2007a: 79–80). For the self to perceive, it cannot float in a vacuum; perception requires a medium. However opaqueness, as we know from experience, is as essential to life as accessibility, otherwise we would be overwhelmed by perceptions and incapable of attention. According to Aldous Huxley (1994), the brain acts as a valve, directing our attention to certain perceptions that enter the body and suppressing others, which are none the less registered by the body. During experiments with various psychotropic drugs, such as peyote, Huxley found that he could see, feel and remember incredible details of colour, texture and movement. However, he found himself totally absorbed by such perceptions and unable to draw himself away from (for example) observing the folds in a piece of cloth. Whether we agree with Huxley's understanding of perception as reception, with the brain controlling that inward flow, is not important for my argument here. Rather, what this example highlights is that our attention is selective and degrees of opacity allow us to have focus, or in my terms 'direction of attention'.

Opaqueness/inaccessibility can be caused by two factors at least: distance or lack of interaction. Although according to ecological phenomenology, all existence is suspended in a continuous, incessantly mutating field of constitutive relationships, we do not perceive all of these interactions. Distance, for one, does not allow us to perceive the entirety of relationships in existence. The effect of interactions in this continuous field can be portrayed as ripples, emanating from an epicentre. As the neighbouring area in the field of existence perceives the ripple, it changes and causes the ripple to travel. However, as with ripples, the energy that is passed on, in reaction to that occurrence, is opposed or diluted. Therefore, the distance between our perceiving selves and that epicentre of occurrence (and every square millionth of a millimetre is thriving with interactions, epicentres of occurrence) dilutes it to our perception, and therefore dilutes our reaction-interaction to things far away. Distance can be understood in terms of magnitude; very small occurrences are also distant in relation humans as perceiving selves.

A refusal to interact can also bring about opaqueness. Gold is the classic example of how communication-interaction can be foreclosed even in close proximity. Gold does not corrode. Its constituents interact with each other, making a 'thing' by their relationship, but they interact far less with other 'things' in the air. Gold does not even interact with oxygen, making oxidisation impossible. In the logic of ecological phenomenology, perception and constitution are different terms for the same process. Perception is not the reception of stimuli, but relational action in the world, which relationships constitute our surfaces and our essences, no matter how provisionally. Other persons are among the other temporally instituted entities within our environment. The consciousness or lifeworld of the 'other' may be opaque to the 'self', due to distance or unresponsive or introvert relations (a lack of interaction). Therefore the concept of 'opaqueness' that phenomenologists employ to describe the lifeworld of others is useful to understand the limits of perception, but cannot be understood as a void between selves.

Social Interaction and Communication Can Only Occur Through 'Sympathetic' Analogy

Rapport does not uphold a solipsistic view; social interaction does occur (Rapport 2003: 59). His claim is only that the connection between people is through forms, habits and symbols which, unless given individual meaning, are empty. Rapport, following Schutz, insists that the only relationship that can exist between the consciousnesses of individuals is based on sympathy: knowing one's own experiences, and by analogy imagining what others feel (Rapport 2003: 256). Ingold (1993), to the contrary, has shown that individuals would not be capable of perceiving anything at all, not even themselves, were they truly as separate as Rapport proposes. Even Schutz trusts that what he calls 'true communication' can occur. There are at least two different ways in which this 'true communication' could come about.

First, there is 'collaborative meaning', an essential part of FoEI's policy-making processes (see Chapters 6, 7 and 8) which, summarily put, is the shared creation of meaning, or meaningful practices. As a concrete illustration of a situation in which collaborative meaning is taking place, recall the occasions when, through conversation, an idea or a plan is formulated through the equal input of two or more people. In these cases, the mutual, collaborative creative process is easily perceptible; the idea or the meaning would not exist were it not for the contribution of more than one person. Apart from this, were collaborative meaning really only an individual adaptation to the form of the habitual actions of others, how would it be possible for people who meet for the first time, and therefore who do not have time to observe a pattern which will allow them to predict action, create something together?

The process of collaborative meaning making can be seen especially when it breaks down. When individuals are intent on forming a common understanding, the process involves considerable negotiation. Interlocutor A presents their position, interlocutor B reacts and adds their own position, interlocutor A reacts in turn and contributes further and so on and so forth. However, if interlocutor B misunderstands, or develops an individual semantic interpretation different from what interlocutor A intended, interlocutor A perceives this in interlocutor B's subsequent contribution to the communicative action, and interlocutor A can correct the situation, saying, 'That's not what I meant'. Many discussions during international FoE meetings took this form. This applies to semantic meaning as well as to action. Ricoeur, for instance, defines performativity as the possibility to correct other people's reactions to (or interpretations of) their actions (Ricoeur 1979).

Of course, Rapport's retort to this could be that the individual will nonetheless create an individual interpretation of the correction. However, if this is still different from what interlocutor A intended, then interlocutor B can be corrected again and again. This is not to say that individuals *never* form different, even contradictory interpretations; on the contrary, it is probably the way most interaction occurs. Tsing (2005) argues that without 'friction' our movements and interactions would not have purchase in the world. However, it is not the only or the necessary type of interaction. When attention is directed towards reaching a common understanding, this *is* achievable. Nowhere is this process more visible than in an apprenticeship. If the apprentice does not understand the individual intention of his master's guidance, he gets it wrong, and the work unravels. The apprentice will try and try again until they have come to a 'doing', a flow of knowing and feeling, close enough to that of their master for the work to hang together (Ingold 2000: 356). The fact that the world has concrete, non-relativistic affordances means that communication also has directly accessible affordances, and when interlocutor B (or the apprentice) pays attention to wrong or unintended affordances/meanings, this can be countered by guiding their attention towards the 'intended' ones.

Second, there is communion, which is distinct from communication (Ingold 2002). When practices are either synchronous or collaborative or both, and when a certain collective rhythm is achieved, people can be in communion with each other. As I understand it, the individual selves merge through the practice, or seen from another perspective, the self becomes transparent and the different individuals engaging in the synchronous practice behave more like a single organism. Wendy James and Judith Aston (2006) show this sort of synchronicity in the grain-pounding activities of women in Uduk speaking villages of Sudan and Ethiopia. The women, in their rhythmic pounding of grain, seek out this communion in their daily work even when not in close proximity. James and Aston showed film of young girls pounding grain, who try to fit into each other's rhythm of pounding and the wider tempo and rhythm of other women pounding in different parts of the village. Such communion is possible because the space between persons and between selves is not a vacuum. When persons apply their energy—their existential power (Rapport 2003)—in the same direction they constitute a collaborative, mutually constitutive wider 'thing' (Latour 2005a, 2003).

Rapport also acknowledges the capacity that humans have to extend their persons. For Rapport, there are no 'because' reasons for the things people do and think. In other words, meaning and action cannot be attributed to the determining force of impersonal structures. Rather, only 'in order to' motives drive people's actions. When responsibility for the inchoate nature of our selves is placed upon an external other, whether 'God, discourse, society' (Rapport 2003: 56) or nature (Latour 2003: 258), Rapport considers this bad faith; a self-deception that not only removes responsibility but also obscures the inherent human capacity for change, imagination and creativity (Rapport: 52, 57).

In the third section of the book, Rapport explores how such existential power confers on beings the capacity to transform one's surroundings according to one's will, by describing the habits of earthworms. Earthworms, which are best adapted to watery environments, manage to dwell in the relatively dry environment of the soil by transforming it, by moisturising it and aerating it with their actions of secretion and burrowing.

Thus, in effect, the soil becomes part of the life-supporting system for the earthworm. This integrated system, which the earthworm creates through its actions, entails that the earthworm has extended its self to continue living in that environment (Rapport 2003: 221–223). Rapport argues that thanks to their existential power, persons are likewise capable of choosing their own life projects and of transforming their environments into extensions of themselves (Rapport 2003: 225).

The notion of existential power explains how social interaction is animated and how social change is possible. Even taken simply from an 'as if' point of view, existential power is a tool of empowerment; it foregrounds choice and reinvigorates moral responsibility. However, closer examination of the illustration of the earthworm's transformative action of the world around it reveals a difficulty. When exploring the relationship of an individual to prior conditions, Rapport (2003: 67) continues to argue that the individual is not determined, rather their relationship to such external conditions is a dialectical one; individuals are in active relationship with, and interpret, external conditions in their own individual way. So too, the earthworm is in an active relationship with the soil around it and interprets it as a habitable place by transforming it as described above. However, if the earthworm's and the person's relation to the environment is an active dialectic, then this environment must also be active. Otherwise, the transformative actions proposed by the person or the earthworm would not be dialectical.

Though the notion of existential power is both explanatory and empowering, the agent—whether individual or earthworm or any other agent—must be seen to exist within the field of mutually constitutive relationships in order for the energies exerted to have any effect in the world. The notion of a field of forces, presented above, allows Rapport's discussion of the expansive self to be contextualised. The constituents of an environment and of a person are mutual partners in bringing about transformations. Through existential power, a person may initiate such transformations in their expansive environment. If there are other persons within this environment, then these others are not just passive vessels—as they are made out to be in Rapport's presentation of the expansive self—but partners, that either facilitate or resist the transformative initiatives of others.

Recalling the description above of the possible actions of vectors, persons may also cooperate and contribute with the expansive, transformative selves of others and thus achieve either communicative action or communion. Indeed, communion and collaborative meaning differ only in emphasis, in that the latter emphasises the mutual creation of something new—a meaning or practice, whereas the former characterises the expanded, although transitional, lively thing that is created when persons invest their energies in the same direction in a shared practice—vectors working side by side.

Systems Are Illusions That Appear in the Aggregation of the Co-Incidental Trajectories of Individual Lifeworlds, and Do Not Constrain Individuals

> The socio-cultural is a concept, existing nowhere but in the minds (the habits of mind and body) of individuals. It does not even exist between individuals. Between individuals are only symbolic forms (words, artefacts, habitual actions), which are inert and empty when not filled with individual meanings.
>
> Rapport (2003: 59)

Quoting John Stuart Mill and subsequently endorsing the statement, Rapport writes: 'The worth of the state in the long run, is the worth of the individuals composing it' (Mill 1972, in Rapport 2003: 65). For Rapport, therefore, systems are like social interaction, empty forms only animated by individual interpretation. Systems, like prior conditions, are 'only what they are experienced to be'. He goes on to say that 'the individual self emerges out of the act of experiencing the world. Ego and environing world are involved in mutual becoming' (*ibid*: 67). In this statement, he would seem to concur with Ingold's dwelling perspective of mutually constitutive relationships and Latour's notion of the progressive composition of the common world. However, were systems, like prior conditions and interaction, only empty forms given meaning by individual interpretation, how could there be 'mutual becoming'? There would be, at most, individual meaning.

As I argued earlier, neither perception nor change would be possible without an ontological precondition. In other words, the continuous world is a condition of being, without which there could be neither perception nor change. Therefore, while retaining Rapport's notion of existential power as well as his emphasis on the personal nature of systems, in the sense that social systems can only be animated through the participation of persons (potentially including non-human persons or things), an alternative explanation for the interplay between individuals and systems needs to be explored.

Recall the notion of reality presented above as a dense, fluid, continuous, albeit punctuated multidimensional field of forces that constitute that existence, in which different magnitudes of forces act in multiple directions. Also, recall the way in which change is brought about, by spaces being opened up for new relations and others being closed off by the various actors in the field. Such spaces are created because the force applied in interaction is vectorial; it has a magnitude and a direction. An actor can apply force in all directions simultaneously, but that would affect the magnitude of the force and create particular affordances. However, like consciousness or attention, most of the time forces are focused in fewer directions. This is where actors have unprotected backs, so to speak.

Rapport argues that everything people do and mean is brought about by 'in order to' motives. Since it is necessary to focus attention or energy to achieve an 'in order to' goal, there are bound to be aspects of daily life that are not given attention—these are the unprotected backs. I suggest that these gaps in our attention are filled by the forces of other actors, whether human or not, animate or not. This is how 'because' causes are possible—how systems and structures can influence our lives—without in any way denying the existence of personal power.

In sum, a reified Durkheimian supra-individual system is not only unlikely but also yields a disempowering understanding of human interaction, especially since it cannot account for creativity, diversity and change. Therefore, systems only exist in so far as does interaction between persons. However, to argue that patterns of behaviour are never carried out involuntarily, and thus that 'because' motives do not exist, is to assume that persons direct their attention in all directions, all the time and with the same intensity, which I maintain is false. Furthermore, individuals often aim to create mutually meaningful practices and invest a lot of energy in ensuring that there are no misunderstandings between the parties. So to maintain, as Rapport does, that collaborative action is nothing beyond empty forms, is just as disempowering as dismissing the creative subject (Rapport 2003: 66). Both individual creativity and collaborative creativity characterise human experience and action in the world.

And finally, if the knowledge of how to attain one's 'in order to' goals within certain social milieux is limited, it can be said that there is distance; while familiarity with actions habitual to certain milieux can be referred to as proximity. When there is distance, it is easy for a person to reify the set of actions that form a system.[9] So far, 'distance' may be the illusion of social systems to which Rapport refers, except that according to Taussig (1992) such reifications, in people's imaginings,[10] can become more powerful than the set of interactions themselves—they become fetishes. But as fetishes they are not unreal, or fictional—and Taussig reminds us of Durkheim's insistence that 'social facts are things'—rather their power comes from the elision of the relations that make up society, for Taussig (*ibid*: 112, 122), or organisations in the case of FoEI. As such, systems may seem to people to act upon them as, and actually have the effects of, supra-human entities the erasure, by distance or lack of familiarity, of the multitude of vectors that compose any *thing*.

Force Fields in This Book

Each of the following seven chapters is developed in relation to the notions of force fields, vectors and direction of attention. In Chapter 3, 'Field Methods, Emplacement and Scale' the different forms of emplacement activists experience, including a sense of the global, point to how multiple topologies are constituted and experienced simultaneously.

In Chapter 4, 'Making History', the books, shelves, room layout and routines that new FoE volunteers are guided through, along with collaboratively imagined and written historical narratives, are some of the forces that are enlisted by FoE activists to stabilise a multitude of relationships and things into their respective FoE groups and, simultaneously, 'Friends of the Earth International'. Sometimes people and other things cooperate, stable 'NGOs' appear and receive funding, activists are called as representatives in public hearings or invited as 'experts' to government think tanks. Websites, offices, photos and reports all attest to the consistent 'thinginess' of FoEI over time. At other times, 'historical' artefacts do not cooperate. Newspaper cuttings and photos begin to fade, crumble or deteriorate due to damp, microscopic organisms, people handling them or heat. Activists leave the organisations, taking their memories with them.

Chapter 5, 'Striving for an Exemplary Life', traces the way FoE activists build their lives into disciplined trajectories. Activists direct their attention towards shaping how they live their lives. Most activists describe their experiences in the countryside as children and youths as constitutive, and yet they continue to learn, change and be constituted by the experiences they undergo all along their life. They are made by and make the landscapes, people and activities they go along with or campaign against.

In Chapter 6, 'Rhythms of Globality', I describe the rhythms and practices, some rhetorical, at FoEI's international meetings. These variously aligned vectors create a sense of belonging among the diverse activists from some seventy countries around the world. Here, vectors are working alongside each other. The belonging engendered at the meetings is part of what allows activists to carry on being present to each other in the daily work they do together by means of communication technologies. In Chapter 7, 'Communication Technologies and Presence', I explore this presence, which also depends on the cooperation of the very particular affordances of electricity and fibre optic cables, as the media with which interpersonal communication can occur.

For FoE activists, their ideals, though imperfect, are part of their experiments in imagining and enacting different possible futures. In Chapter 8, I contrast the ideals upheld in FoEI documents and discourse with their practices. Here, ideas are vectors that are future-oriented; they emerge as tools with which the activists experiment with different organisational structures and decision-making processes.

Notes

1 It is beyond the scope of this book to present the arguments against all these other paradigms. Further on in the chapter I make a close argument against one other frame of thought, inscribed in the work of Rapport (2003). However, for arguments against these other paradigms, see Ingold (1992, 1993, 1996, 1999, 2000a, 2000b, 2006, 2008a), Latour (2003, 2005), Haraway (1991, 2003), Farnell (2000), Asad (1979), Pfaffenberger (1992), and Ratto (2009).

2 A key text on social constructivism is Berger and Luckmann (1971). See Eeva Berglund (1998) for a bibliography of 'social constructionism'. She points out how the various labels 'constructivism' or 'constructionism' are not necessarily

self-designated. Berglund includes Latour in her references. Indeed many others do not differentiate Latour from others in social constructivism, STS or ANT. However it is essential for this synthesis to note that Latour, from *Politics of Nature* onward, clearly indicates that the social cannot be understood in Durhkeimian terms, as a supra-natural domain, and rather talks of 'the progressive common constitution of the common world' by all sorts of matters of concern (2003). He explains this explicitly in *Reassembling the Social*. In addition, see John Shotter (2010) for an approach to social constructivism that explores human persons in relation to movement and ecology.

3 Haraway's (1991: 190) arguments about perception relate more specifically to the necessity to see from different positions. However, she also criticises arguments that propose 'some abstract version of existence' (*ibid*: 153).

4 Power and agency has been envisaged as force fields by many scholars. Not least by Walter Benjamin, following his work Althusser (Martin 1993) and also by Michel Foucault (Kusch 1991). Although I acknowledge the prevalence of this metaphor in the literature, and I must admit to not being expert in their work, for various reasons I develop my own understanding of force fields. Neither Benjamin, Althusser nor Foucault adopt the approach to force fields where they are both metaphors and metonyms. Their interest was not to account for non-human factors. For more details, see Gatt 2011, Appendix 1.

5 No anthropomorphism intended—recognition may be the type of intentionality we know in humans, but also the reaction to temperature of a thermostat (David Chalmers 1996), or the speech of non-human things in the way they act/react in different circumstances, which Latour (2003) calls *speech impedimenta*.

6 For a detailed exposition of the difference between the notion of the network as connecting ready-made entities and relations that participate *alongly* in interwoven unfolding lifeways, see Ingold (2007a: Chapter 3).

7 Similarly, a methodological struggle I faced in fieldwork was: if for environmentalists everything is connected, what should I focus on to build an account of their work and lives? I found that the solution was to follow a pattern of 'serial closures', that also helped me contribute concretely to the FoE groups I worked with (Gatt 2010).

8 Also see Lave (1988) and Clark (2001) for other approaches to cognition that does not separate the mind from the rest of the person-organism.

9 Consider Gupta's (1995) article about a state official in Northern India, on the experience of, and wayfaring with, imaginations of the 'state'.

10 Experiences of imagined perceptions are as real as those of tangible perceptions. I argue that imagination, together with thought, are equal tools of mutually constitutive relations with the environment. Imagination is alert to certain affordances that other aspects of perception are not. Isabel Allende advises that, 'it isn't sensible to rely solely on reason and on our limited senses; after all other tools of perception exist, like instinct, imagination, dreams, emotions and intuition.' (Allende 2004: 73).

3 Field Methods, Emplacement and Scale
Where Is FoEI?

The fifth point of the FoEI Mission, reads:

> To engage in vibrant campaigns, raise awareness, mobilise people and build alliances with diverse movements, linking grassroots, national and global struggles.

Thus, one of the aims of the International Federation is to 'link' activists' efforts. These efforts are imagined as happening in many places and on diverging scales. The International Federation exists to concert the activism spread around the world. As such, FoEI is not produced in one site or another, but in many simultaneously. In addition, the environmental issues around which the activists struggle are also considered to stretch across different places and scales. Multi-sited methodology would be an obvious choice for this project that set out to understand the constitution and maintenance of FoEI.

Multi-sited methodology, famously captured in a review by George Marcus (1995), is a form of research design that focuses on connections between and across 'sites' of cultural production that are not necessarily geographically contiguous. Marcus (1995) refers to this as 'following' the subject in a number of possible ways. In fact, a key argument for using multi-sited methods is that of ' "responsiveness" to the anthropological subject' (Falzon 2004: 13). Okely (1992) describes responsiveness in anthropological methods as an active and searching openness about anything and everything. She describes such responsive methods by a term that means available-ness: *disponibilité* (*ibid*). Rather than multi-sited methodology, I adopted *disponibilité* as a methodology for this research project and allowed it to guide my understanding of the 'sites' of fieldwork. In this chapter, I trace the course of my fieldwork with FoEI and the different methods I used and how these were responsive, *disponible*, to the situations that arose through fieldwork. I expound on the implications for anthropological approaches when *disponibilité* informs understandings of place, and therefore how we theorise the 'sites' where we do fieldwork.

According to Marcus (1995), the connections across sites explored in multi-sited fieldwork are understood to derive from world-spanning systems or flows, such as colonialism and capitalism. However, multi-sited method-ology is different to previous studies of world systems in anthropology. For these the world system provided the context for studies in particular locales (Marcus *ibid*). Rather, multi-sited fieldwork juxtaposes ethnographic field-work from different nodes of such systems, thus providing, in effect, ethnog-raphies not only of sites within but also studies of those systems (*ibid*). By carrying out fieldwork with differently positioned nodes within such world flows and systems, Marcus (*ibid*) argues, the multi-sited studies he reviews collapse the distinction between lifeworld and system and thus provide ethnographies of the production of those very systems. Candea (2007) has argued that there is a degree of holism implicit in the project of multi-sited fieldwork, in that a defined whole is imagined to exist even if Marcus rejects holism in that a world system is impossible to capture in any number of studies. The question of holism and world-spanning systems is relevant to an understanding of FoEI.

FoEI is an organisation, whose members claim that it is a coherent entity. Therefore, a multi-sited study of FoEI could more realistically include dif-ferent positions of 'the system', if the system is FoEI. Therefore, informed by multi-sited methodology, the research design for this project included three nodes, differently positioned, *within* FoEI. However, heeding Candea's caution against reproducing an imagined bounded whole and following the methodology of *disponibilité* three things became clear. First, that FoEI as a bounded entity only exists and has effect as an entity when specific vectors are at work. Second, I will suggest that FoEI activists' emplacements in the world do not fit neatly with Marcus's understanding of 'sites', due to the spatial assumptions inherent in his logic. Finally, there is more to the activ-ists' emplacement than *ideology* of scale and place.

During the course of my fieldwork with FoEI, it became apparent that the sites where FoEI activists interact with each other around the world were not simply different 'spatial configurations' (Marcus 1995). The sites of engagement, which produce what it means to be a FoEI activist, differ from those of archetypal 'single-sited' fieldwork not only in the sense that they are geographically located differently to those of other social groups. By working with FoE activists, it became apparent that different types of topologies emerge as a result of their daily practices. These practices, some particular to FoE activists, others not, bring them into constitutive interac-tions with things such as telephones, maps, online services such as Skype and Google Earth, email and instant messaging (IM); with offices and their architecture, tables and books. Those participating in the activities, includ-ing the people and these things, but also the medium for such activities (air, electricity or fibre optic cable) form not only the constraints of the activity but they participate in generating the actual quality of that place. In other words, 'places' are created by practices not *only* in the sense that those practices give importance or significance to specific locales. Those practices and actors with particular affordances who participate, actually generate the material quality of place as well. Therefore, different practices create different qualities of place, their emplacement or different topologies. In fact, multiple topologies have been observed in conventional sites, albeit as pertaining to different groups of people (see Lund 2005). Whereas I observed that different practices with different actors, human or non-human, created different topologies simultaneously for those involved. As a result, my fieldwork can be described neither as multi-sited nor single-sited. Rather, the fieldwork was driven, in part, by attention to the significant participants, human or non-human, that co-produced different topologies.

Tracing the Course of Fieldwork

Tracing the continually changing course of my fieldwork will illustrate how *disponibilité* informed the preparation and carrying out of my fieldwork in practice. Overall, the fieldwork for my doctoral research project comprised six months of participant observation with FoE Brazil, five months with the FoE International Secretariat in Amsterdam, and six months with FoE Malta. As part of my doctoral fieldwork, I also participated in nine international FoE meetings between 2003 and 2007. Throughout my fieldwork, I kept up email participant observation with FoE Malta, and later with FoEI email discussion groups. I also draw on my experience from the period between February 2003 and September 2005, during which I engaged with FoE Malta as an activist (part employee, part volunteer).

In February 2003, I began working with Friends of the Earth Malta (FoE Malta). I was employed as the coordinator for an applied anthropology community project. As an employee and a volunteer with FoE Malta my understanding of who the activists in FoE Malta were, what they did and what they had done in the past, their areas of interest and the principles they adhered to, and what the wider FoE network was, became clearer gradually. Anthony, the current coordinator of FoE Malta, recently described his experience of getting to know Friends of the Earth (FoE) in a very similar manner. Julian Manduca (his actual name), one of the founders of FoE Malta, guided Anthony in getting to know FoE and FoE work. He told me: "I first met Julian. Then at the next meeting, I met a few more people. And after a year I had slowly pieced together the picture of the organisation, like a jigsaw puzzle" (from fieldnotes 12th June 2004). To a great extent, my fieldwork followed a similar course. In what seemed like fragmented stages, by meeting more and different people, by engaging in discussions, by participating in email groups and not least by reading the literature on environmentalism, I came to know more about how this thing 'Friends of the Earth' might be brought about.

As I went along, my research strategy took rather divergent paths with the various places I visited, the different people I worked with and the different organisational contexts within which they engaged. After having worked with FoE Malta for two and a half years, in 2005, I was given the opportunity to carry out doctoral research with FoEI. At this stage, the plan was to carry out fieldwork 'at home' with FoE Malta, with FoE Europe (FoEE) in Brussels, a regional FoE body, and finally with the International Secretariat (IS) of FoEI in Amsterdam. The strategy reflected what I had learned about FoE from being a part of FoE Malta. For FoE Malta, the key points of reference in FoEI were the IS and the European regional body. I did a further three months of fieldwork with FoE Malta, after which I accompanied Anthony to two FoE Europe meetings. During the second of these meetings, it became clear that the FoE activists perceived considerable differences between the 'weight of responsibility' for environmental degradation of the European region on the one hand and of the African, Latin America and Caribbean (ATALC) and Asia-Pacific-Oceania (APO) regional groupings on the other. This perceived division within FoEI echoed what is found in the literature on development and environmentalism, which describes a distinction between countries along a North/South divide. In order to explore the grounded experiences within and around these distinctions, and following the advice of my supervisor, I changed my fieldwork plan.

In 2006, my plan retained FoE Malta and the FoEI Secretariat in Amsterdam, but instead of the FoEE office in Brussels I would spend six months with Núcleo Amigos da Terra (NAT/FoE Brazil), considered a 'Southern' FoE group. In comparison with FoE Malta activists, FoE Brazil activists refer daily to other FoEI activists, projects and campaigns. Significantly, the majority of FoE Brazil activists are employed, while FoE Malta activists are all volunteers, except one. Finally the continental difference in geographical size between the respective nation-states of Brazil and Malta involved radically different types of 'following the people' (Marcus 1995) in the two countries.

In Malta, I lived at home with my parents, and met FoE Malta activists daily in email discussions, meetings in cafes and in people's homes, including meetings at my own home. Fieldwork in Malta also included going to protests organised by other environmentalist NGOs, organising and going to a number of 'educational' events, and participating in the meetings of the 'Nature group' (an informal grouping of ENGOs). I also spent a lot of time with Anthony, the coordinator of FoE Malta, accompanying him on his weekly hikes across the Maltese and Gozitan coast and down to hidden rocky coves.

Fieldwork in Brazil involved a considerably greater volume of emails and time in the office. However, the most obvious difference was accompanying FoE Brazil activists to meetings in Porto Alegre, in the state of Rio Grande do Sul, and to meetings in other states of Brazil. I travelled with FoE Brazil employees for meetings and protests to the northeast to Porto Velho, north to Teresina and to the centre of Brazil to Belo Horizonte, but also closer, to Santa Katarina. With some activists, I visited their family eight hours away by bus, and I went with FoE Brazil activists on their holidays to the beach and on a hike in the Atlantic rainforest. In Brazil, I shared a flat with a FoE Brazil volunteer from Australia. During my time in Brazil, I learned that the FoEI annual meetings and the FoE IS were not the only contexts in which activists from FoE groups worked together. There were many campaign working groups, with their own email groups and regular meetings—in fact, campaign working groups try to hold at least two meetings a year. To carry out fieldwork in Brazil I had to learn to speak Portuguese, which was facilitated, in my case, by knowledge of other Romance languages. As Marcus (1995: 101) imagined, fieldwork had to be multi-lingual as well as multi-sited.

Once I got to Amsterdam in 2007, to spend five months there, I was also hoping to accompany the different activists in the Secretariat to these campaign meetings. However, accompanying the activists from the Secretariat proved to be an entirely different matter from what it was in Brazil. The campaign meetings were held in Cameroon, South Africa, Brussels, Bali and one in Amsterdam. I could only attend the one in Amsterdam. I was permitted to attend both the FoEI Gender working group meeting in Amsterdam and the annual FoEI meeting in Swaziland, only by virtue of being a representative of FoE Malta. Other researchers who had asked permission to observe the meetings were turned down. As a result, fieldwork with the IS turned out to be primarily based in Amsterdam itself, spending a lot of time in the office, attending a few 'social' gatherings, carrying out archival research and making some home visits, essentially to carry out life history interviews. The office managers preferred that I did not live with any of the FoEI staff. Sharing a home with a researcher was perceived to be a continuation of work and the managers felt their staff should not have to take work home. Instead, I ended up living with an ex-FoEI volunteer. While in Brazil I took many photos;[1] in Amsterdam, by contrast, it was clear that people were not comfortable with pictures being taken in the office. I also have few photographs from the first two periods of fieldwork in Malta, since there were not many face-to-face meetings and most interaction happened by email. I carried out life history interviews with activists from FoE Brazil, FoE Malta and the IS.

A central aspect of the fieldwork with both FoE Brazil and the FoE IS was the fact that the activists were extremely busy, and their work was characterised mainly by reading and writing documents, email discussions and other computer-based work. I could only ask for details and explanations, as is customary (Wikan 1993: 201), during lunchbreaks or especially when moving between meetings—on foot, by car or plane in Brazil, on foot or on our bicycles in Amsterdam. Researching the particular practices related to FoEI, then, meant that I had to participate in meetings and email discussions, and to devise alternative methods. In Brazil, one of these alternatives was the work maps that I and a FoE Brazil activist drew together on card sheets.

The lack of 'conversational' engagement during fieldwork in Brazil and Amsterdam—because people were so busy at their computers—provided the opportunity to observe other aspects of the activists' practices and engagement with their environments. For instance, I paid particular attention to the different daily rhythms, the use of their tools—their chairs, tables, books, computer keyboards and other computer hardware—as well as the spatial relations within the offices and meeting places. In particular, these observations informed my interpretations of history and growth in FoE groups (Chapter 4), of shared rhythms at FoEI international meetings (Chapter 6) and of the activists' use of communications technologies (Chapter 7).

Eventually I returned to Malta for another three months, by which time there were considerably more face-to-face meetings, and more demonstrations—as other environmentalist groups organised more frequent public events. Though FoE Malta activists did not participate in the organisation of these demonstrations, a handful of FoE Malta activists carrying the FoE Malta name-flag and placards would always go. In addition, I carried out more life history interviews and spent brief periods helping out in the new FoE Malta office. Each time I returned 'home', the practices of FoE Malta had changed. However, my fieldwork with FoE Malta carried on while I was in Brazil and in Amsterdam. I kept in touch with FoE Malta on a regular basis, first via email, then also with Skype. I was more active in regular communications with other FoE Malta activists while I was in Brazil and then Amsterdam, than previously when I had to focus on the community project.

Even had I not deliberately adopted *disponibilité* as a stance towards field methods, each FoE context required different methods. More often than not the situation made it clear which methods could *not* be used. The different FoE activists' work practices and the wider context within which their work was situated vis-à-vis FoEI differed in ways that required different methods. However, the activists I engaged with all imagined themselves to be part of the FoEI network. The very differences of their experiences are considered the threads that constitute the sum total of FoEI. Their different perceptions are not 'reterritorialized' (Inda and Rosaldo 2002), or in other words imported and reinterpreted by different FoE national member groups. Rather their experiences of and contributions to FoEI are considered to be constitutive of FoEI itself.

Multi-Sited Practice/Theory as Enabling

At all stages of my research, discussion of multi-sited[2] methodology provided a provocative sounding board against which to identify resonances and differences within the milieu of FoEI. It was also pivotal in stimulating my sensitivity towards 'sites' and emplacement. The work on multi-sited methodology was enabling in three particular ways.

First, it suggested particular methods with which to get started. In the light of FoEI's organisational structure, it was necessary to explore the relations *between* FoEI member groups to understand the constitution of the organisation. Marcus's (1995: 106) review offered the practical suggestion to follow some people, to follow their lives and biographies (*ibid*: 109). The ethnographies of Garsten (1994) and Falzon (2004) suggested that for various reasons, including financial, three was a practical number of locales to work in. I therefore chose to carry out fieldwork with three FoE groupings. These methods were a contingent guideline by which to go about trying to understand how FoEI 'hangs together' (Hannerz 1996: 4).

Second, discussions of the binaries global/local and system/lifeworld and nation-states (stimulated by multi-sited practice/theory) were particularly useful in formulating questions of relevance to FoEI due to its particular ideology and structure. FoEI activists seem to live a contradiction. On the one hand, the prevailing environmentalist ideology in FoEI recognises the Earth as a continuous, interconnected and interdependent world. On the other, its membership structure, divided along country borders, implicitly endorses the logic of the nation-state, which does not incorporate the continuity and interdependence of environments as understood in environmentalist terms. The points that Marcus (1995) raises about the dichotomies lifeworld/ system and global/local resonate with the experiences of FoEI activists, who also deal with these dichotomies.

Marcus's (1998: 40) critique of the dichotomy lifeworld/system is that it homogenises the notion of the 'system', while assuming that 'local' (subaltern) lifeworlds are diverse. Multi-sited research offers a possible way to collapse the dichotomy, as Lash and Urry's (cited in Marcus 1995: 50) work does, by providing an 'ethnography of complex connections' between places that are produced in/of (and I would add produce) a world system. Haraway (1991: 190) proposes that since all knowledge is situated—I discuss her specific definition of 'situatedness' below—in order to 'see well' it is necessary to explore different situated knowledges. My choice to carry out fieldwork with three groupings of FoEI was designed to explore different types of situated knowledges within the FoEI Federation.

Third, discussion of the constraints of multi-sited methodology offered a point of departure from which to ask further empirical and theoretical questions. The most common concern raised about multi-sited methodology is loss of depth. The compelling response to the problem of depth is that a multi-sited project is not, by definition, comparative (Hannerz 1996: 237). Falzon agrees with Hannerz, pointing out that 'the notion of a 'complete ethnography' was always something of a myth' (Falzon 2004, 21). Falzon (*ibid*) shows how his ethnography of the Sindhi diaspora focuses on one important aspect of the lives of the Sindhis he researched: the translocal aspect. The practice/theory of multi-sited methodology emphasises this point. It is the connections and the relations across what are considered at times to be different sites that are investigated (Marcus 1995, 1998), creating in effect a single research project.

Appadurai (2005) proposes extending Anderson's idea of imagined communities to 'imagined worlds' to understand such relations. In a similar vein, Rouse (2002) describes the lifeways of Mexican migrants to the United

States as a 'community dispersed in space'. Rouse (ibid) elucidates this spread-out community through migrants' travels back and forth between Mexico and the United States, the fact that they equip their children with familiarity with both contexts, but also through their continued participation in social obligations and relations at a distance primarily by means of the telephone. The emphasis of multi-sited practice/theory on relations that do not necessarily depend on continual face-to-face interaction has achieved two things: it has destabilised the emphasis on face-to-face participant observation and as some claim its inherent bias towards bounded sites (Gupta and Ferguson 1997; Amit 2000; Coleman and Collins 2006); and it draws ethnographic attention to *how* such relations are lived (Garsten 1994; Rouse 2002; Falzon 2004). Finally, the practice/theory of multi-sited methodology equipped me with an indication of *what* to pay attention to during fieldwork in order to explore, among other things, the non-face-to-face relations—if there were any—that kept FoEI 'hanging together'—if it did. This education of attention[3] allowed me to perceive some discrepancies between the lived experience of such relations and the ways it is perceived in current anthropological discussion.

'The Field', 'Location' and 'Site'

The context for recent discussions of the 'field' and 'location' in anthropology lies in a practice of ethnography that is in some way multi-sited (Gupta and Ferguson 1997; Amit 2000; Coleman and Collins 2006). Anthropologists engaged in such practices encounter people whose lifeways are constituted to a significant measure by movement and flow—either their own or that of images, ideologies, commodities or capital (Inda and Rosaldo 2002). The life experiences of such people challenge the immobility and boundedness demanded by the view of the world as a mosaic of geographically delineated cultural wholes (*ibid*: 11). Anthropology remains committed to developing theory drawn from empirical findings, or that at least does not contradict the experience of informants (Coleman and Collins 2006; Tsing 2002; Wikan 1993). As a result, multi-sited ethnography has reinvigorated the critique of bounded understandings of cultural difference and a focus on 'sites' (Marcus 1995: 112).

Interrogation of the notion of anthropological 'field sites' has brought into question the limits of ethnography. The Malinowskian ideal of fieldwork—fieldwork based on prolonged participant observation in a relatively small, geographically delineated place—had become the hallmark of anthropology, distinguishing it from other disciplines, such as history and cultural studies (Gupta and Ferguson 1997: 3). Crucially, Gupta and Ferguson (*ibid*, 12) show how the mosaic view of cultures, though rejected as a theory, is nevertheless perpetuated in the notion of, and adherence to, the Malinowskian ideal. They examine a number of aspects of how fieldwork is carried out and the discourse around it that maintains the notion of boundedness. These include the tropes of entering and exiting the field, the hierarchy of field sites and the emphasis on face-to-face participant observation.

Gupta and Ferguson (*ibid*) point out how discussion of fieldwork methods inevitably leads to discussion about the discipline of anthropology. However, along with Coleman and Collins (2006), they denaturalise the imperative of Malinowskian fieldwork by tracing the history of fieldwork methodology highlighting other methodologies, orthodox and heterodox, in anthropology's past. They demonstrate how Malinowskian fieldwork has not always defined what anthropology is and does, thereby enabling a wider discussion of methodology in anthropology.

Almost ten years later, Coleman and Collins's (2006) introductory chapter, addressing very similar issues as Gupta and Ferguson's (1997), seems to need reiteration. This suggests, possibly, that the Malinowskian fieldwork ideal is still hegemonic. In addition to deconstructing the centrality of Malinowskian fieldwork, Coleman and Collins (2006) trace how boundaries, embodied in the demarcation of a single geographically delimited field site, have been useful for cultural comparison in anthropology (*ibid*: 8). Synecdoche—where a part is taken to represent the whole—and bounded sites co-existed in creating the anthropological subject, setting up the cultural difference in which anthropology is supposed to specialise (*ibid*). However, with 'globalisation' the synecdoche local/global does not work, since the global has no lines of division along which comparisons can be drawn (*ibid*).[4] Thus, while for Gupta and Ferguson (1997), multi-sited practice/theory challenges the limits of Malinowskian fieldwork method by destabilising a dependence on the archetypal 'natural laboratory' of the bounded field site, Coleman and Collins (2006) draw attention to the perceived threat to the anthropological context (created by regional comparison) that 'globalisation' studies usher in.

Neither Gupta and Ferguson, nor Coleman and Collins, accept that the challenges to Malinowskian fieldwork signal the demise of ethnography and they offer, as does Amit (2000), very similar proposals for a reinterpretation of fieldwork. Gupta and Ferguson (1997) argue that the anthropological 'site' is a social construction. Thus, in order for contradictions between theory and practice to be avoided, what needs to be explored is the ways people construct locations (*ibid*: 35–39). While they do not propose any single alternative methodology to single-sited fieldwork, they do propose 'shifting locations', as opposed to 'bounded fields' as a way to define where and about what anthropological fieldwork is done (*ibid*).

Amit (2000: 6), criticizing the persisting image of the solitary fieldworker, argues that in the current atmosphere of reflexivity ethnographic texts need not omit fieldworkers' personal relations. She argues that the 'field' is constructed by disentangling and describing certain relationships and not others, to provide one contextualisation and not another out of multiple possibilities (*ibid*: 6).

What Coleman and Collins (2006) add to the notion that the field as a social construction is that anthropology remains committed to investigating the lifeways of its informants. As such, fieldwork is emergent from both the agency of the researcher *and* the serendipity of the unexpected (*ibid*: 7). They suggest 'performing the field' as a way to liberate fieldwork from 'the purely spatial metaphor' (*ibid*: 12) while maintaining the idea of the mutual creation of ethnographic knowledge by researcher and informants. In resonance with ethnographic experience of mobility and flow, these writers claim they are reacting against the immobility and boundedness of single-sited fieldwork. The solutions they propose—'shifting locations' (Gupta and Ferguson), 'the constructed field site' (Amit) and 'performing the field' (Coleman and Collins)—emphasise the social construction of place, location and the field site.

In summary, multi-sited practice/theory has stimulated a productive discussion about 'field site' and 'location'. This discussion, while describing current trends, also brings them and the issues they raise to mainstream attention, allowing them to be more widely discussed. These analyses pave the way for methodological change in orthodox understandings of fieldwork as well as addressing the feared demise of ethnography or anthropology that such a change would purportedly bring. The proposals that liberate fieldwork from geographical boundedness enable anthropologists to bring their research to different or new areas of human experience, as is the case of my research with FoEI. Additionally, reflexive analyses of fieldwork also encourage further investigation and debate. One point, in particular, calls out to be debated: namely, that regarding location and situation (including the location of the field) as socially constructed. This has the effect of detaching sociality from emplaced lived experience.

The detachment of the 'social person' from the human organism and the non-human environment, in other words from their lived emplacement, is explicit in most of the literature discussed above. Gupta and Ferguson call for location to be understood as social construction, yet they do not mention the environments in conjunction with which such constructions are made. Coleman and Collins explicitly state their position that anthropology is about social relations, agreeing with Hannerz that it is only derivatively about place. Marcus (1998: 45) suggests that multi-sited fieldwork transcends the local/global dichotomy by 'experimenting beyond place-focused . . . ethnography'. The implication is that the non-human aspects in particular situations have no bearing on the relations that unfold there.

In order to understand how social groups cohere, there seems to be a shift from one extreme to the other, from immobility and bounded cultures to dis-emplaced mobility and flow. On the one hand, the shift has *enabled* the extension of anthropological research into areas of human experience that the bounded site methodology did not allow. Yet on the other, the imputed detachment of people's social relations from specific environmental, emplaced relations in order to allow for mobility and flow, perpetuates the same assumptions about the human constitution that established single-sited fieldwork in the first place.

The Detachment of the Social from Emplacement

Malinowskian fieldwork was part of the same movement as fieldwork in natural history (Gupta and Ferguson 1997: 6–8, Giddens 2009: 163). The movement considered that empirical information collected in the 'natural environment' was the most reliable. Another assumption inherent in this movement was that so-called 'primitive' societies were closer to nature. 'Complex societies' could not be studied in their 'natural environment' because civilisation had the effect of detaching people from 'nature'.

Durkheim expounds on the variable proximity of societies to 'nature' in *The Division of Labour* (Biernacki and Jordan 2002). He argues that if the practices of a group of people, provided the group is not too extensive, are attached to a concrete landscape, say to a river or a forest, then individuals will automatically follow shared norms and beliefs (*ibid*: 136). When, however, people interact with others from different places (with different rivers and forests) then their relationship to the environment becomes abstract. Collectively they do not relate to a particular forest, but to the abstract notion of forest (*ibid*). Through this abstraction, social beliefs, norms and relations form a society that is *sui generis* and follows its own internal laws not influenced in any way by the environment (Hatch 1973: 188, 175). The social connections between people are therefore suspended in a social space, and rather than relating to the environment, collective beliefs *transfigure* the non-human world (*ibid*: 203).

As Biernacki and Jordan (2002: 136) show, the abstraction of social space from ecological relations is understood in the sociology of Durkheim to be an inevitable outcome of history. The more complex the division of labour in a society the more its constitutive social relations become detached from the environment. Henri Lefebvre argues likewise, that with the onset of modernity people's relationship with lived places is increasingly dominated by abstract space. In fact, Lefebvre (1991: 123) admonishes anthropologists for investigating 'social space' in non-modern societies, where it is not to be found. This ontology, shared by Durkheim and Lefebvre, separates those people who have 'social relations' and who inhabit 'social positions' or 'social space', and are therefore not influenced by the 'environment', from those who have 'culture' and who need to be studied in their 'natural environments'.

The practice/theory of multi-sited methodology follows Durkheim and Lefebvre in associating modernity with detachment from 'place', that is, from constitutive relations with a non-human environment. The progression that is envisaged goes from pre-modern (close relations with the environment) to modern (social relations abstracted from the natural environment). This is a shift from understanding people as inhabiting places to their inhabiting abstract space. Places are 'loci defined by the layout of a setting *for practice*' (Biernacki and Jordan 2002: 134 italics added); in other words they are con-stituted by the multiple fields of relations people work through and with, in their daily lives. On the other hand space is abstract and conceived as a naturally given, grid-like platform for human conduct . . . reducible (or already reduced) to a formal (that is, empty) schema or grid. It establishes a featureless, purely extensional arena for action apart from its occupation by an embodied agent.

(*ibid*: 133).

When, for instance, Marcus suggests that multi-sited fieldwork can trace relations that are on a 'differently configured spatial canvas', the implicit notion is that of isotropic (the same everywhere) space, a background canvas upon which social relations can have different arrangements. In this view 'the landscape itself is rendered uniform; it is reduced to 'space', a vacuum to the plenum of culture' (Ingold 1993: 226).

Most, if not all of the multi-sited, transnational and globalisation studies reviewed by Marcus (1995), Hannerz (1996) and Inda and Rosaldo (2002) explore in some way aspects related to capitalism or colonialism. Moreover, although it is always qualified, the recent interest in mobility and flow is explained by the development of communication and transport technology (Inda and Rosaldo 2002; Erisken 2003). Transport physically moves people and commodities from place to place; communication technology moves capital, ideologies, images, emails and voice. Due to the perceived ubiquity of the abstracting and detaching tools of modernity, people's relations with the environment are also understood to be everywhere abstracted, mediated by social positions. Thus, the understanding that social relations are separate from people's relations to places, in other words from their environments, implicitly endorses the ontology of natural history in which Malinowksian fieldwork was established—except that now, no 'natural environments' exist and everything is only about 'social' location.

Haraway argues that infinitely mobile vision is a disembodiment from situated vision and as such is an extension of the 'god-trick' (1991: 188–189). Visualizing technologies have turned the illusionary 'god trick' of seeing everything from nowhere into seeing everything from everywhere and, moreover, insert this into everyday practice (*ibid*). Therefore, the proposal to focus entirely on social relations as a solution to the immobility and boundedness of single-sited fieldwork perpetuates the 'god trick' by failing to explore the particular environmental relations in conjunction with which mobility and flow come about. I am not, of course, trying to revive the cultural ecology of the 1960s. It has been amply shown that the same non-human environment can afford many different ways of life. However, the alternative to the idea of geographically bound cultures is not necessarily that all landscapes are 'socially constructed' in the sense that non-human or non-cognitive things in the world are of a different domain to the 'social' and do not participate in the constitution of 'places'. An alternative that does not detach the 'social' from the non-human, non-cognitive world is to be found in the ecological phenomenology discussed in Chapter 2.

In a parallel discussion in anthropology, Ingold (2000b: 215) has argued that 'the ancient division of the human being into two parts or aspects, social subject (person) and biological object (organism)' has also been maintained in scholarship on landscape and in that on embodiment. The Cartesian mind/body separation, though discarded as a theory, has been perpetuated by the continued lack of interest in the ecological relations of the organism-person (*ibid*). To overcome this separation, he suggests that researchers should direct their attention to the organism-person who participates in a field of ever-unfolding relationships with the manifold constituents of their environment (2000b: 216). Lee and Ingold (2006) provide an example of sociality understood as constitutive relationships between those manifolds participants in an environment, and not as a discontinuous cognitive domain. In a research project about walking in Aberdeen, they show how whole persons and the environments they walk through co-produce routes of sociality (*ibid*: 72). Sociality, here, does not divide human social personae from the physical world and non-human aspects of an environment, but it indicates the mutual constitution of that 'route', that 'way of life' of the humans they worked with, the road, the weather, other people and so on. What is brought into focus is not only how the non-human aspects of an environment affect the persons making their way through them throughout their lives. This ecological approach allows an understanding of how particular shifting environments are produced and affect anything living within them, while seamlessly including those aspects previously cordoned off as 'social', such as memories, semantics and symbolism.

Apart from perpetuating a division between mind and body, the assumption that non-human constituents in an environment are detached from 'social' relations has a specific implication for research. The idea that places are 'socially' constructed but 'physically' located in isotropic space projects onto the lived experience of research subjects that the non-human ground is homogeneous, when in reality this is far from the case. In the following section, I attempt to convey some of this diversity through different topologies that FoEI activists live through. These do not 'transcend' or 'transfigure' space through 'socially' constructed meaningful places, but are made possible and co-produced by what the manifold non-human as well as human constituents of their environments afford.

Neither Multi-Sited nor Single-Sited; Rather, Topological Multiplicity

My fieldwork with FoE Malta carried on while I was in Porto Alegre and in Amsterdam. I kept in touch with FoE Malta activists on an almost daily basis via email and Skype. I was active in FoE Malta email discussions, taking decisions, proposing ideas, editing documents and press releases. Though FoE Malta and FoE Brazil had regular contact with the International Secretariat, they did not at the time directly cooperate on any projects together. On the one hand, FoE Malta and FoE Brazil activists consider themselves constituent actors of FoEI, and they feel that any work done by FoEI in any part of the world is part of their FoEI too. Thus, my fieldwork was single sited in the sense that I was working with a circumscribed group bound by thick social relations, albeit dispersed in a much larger geographical area than the traditional single site. On the other hand, certain distinct historical, geographical and ecological aspects of the places FoE activists inhabit—including the nation-state—are the basis upon which the FoEI structure is modelled. In this sense, the fieldwork was multi-sited since the geographical distinction between sites is a constituent part of their very relations. The spatial assumptions implicit in the terms used in anthropological methods did not match the FoE activists' experiences of space. I came to realise that neither 'multi-sited' nor 'single-sited' adequately described the places in which I was doing fieldwork.

FoE activists experience, inhabit and co-produce at least three types of places *simultaneously*. They consider the whole world to be a single site—indeed the whole universe, according to those FoE activists who are also amateur astronomers—a belief common to the environmentalist ideology of the Gaia hypothesis (Milton 2002: 31). FoE activists will recognise and act upon a mutual recognition of belonging to FoEI regardless of the explicit diversity of the individual FoE activists and their diverging ways of being environmentalists. At the same time, activists engage in certain practices and hold certain beliefs that configure their lived emplacement into bounded sites—divided along national lines, divided into different ecosystems with fuzzy borders, and often divided according to political positions or group affiliation. Each of these very different ways of experiencing places and the social relations they afford is emphasised, depending on the activity at hand, but they are always at play even if not focused on at all times.

The different experiences of emplacement are contemporaneous aspects of how FoEI activists constitute and experience being part of FoEI. As I mentioned previously, these different types of emplacement appear to rest on what Tsing (2005) calls 'ideologies of scale'. However, as Strathern (1991: xiv) has argued, 'scale' belongs to a particular intellectual tradition, and with the implication that the particular ways FoEI activists construct 'scale' itself need further examination. In fact, the diverse experiences of emplacement that I encountered with FoEI activists resonate closely with the different topologies set out by Mol and Law (1994). For each topology, scale is an entirely different matter.

The notion of topology is useful in discussing FoEI activists' emplacement for various reasons. First, as the study of how place is experienced, topology calls on us to question environmental relations, rather than assuming what places and sites are and how they are constituted or experienced. Second, as Biernacki and Jordan (2002) show, 'abstract space' may still be a part of people's experience and, as such, is more than just a concept of analysis. It also has to be recognised in empirical investigation. Similarly, the topologies

Mol and Law describe include attention to how 'abstract space' is enacted in what they call 'regional topology'. Third, the notion of topology also avoids further dichotomisation of place and space, which would make it difficult to understand how both can be simultaneously part of experience without contradiction. Although Mol and Law (1994) also separate the social realm from the lived environment, stating that their topologies are only metaphorical of social situations, their development of the notion of topology, as well as the actual ethnography they write, *draws attention to environmental relations*.

Through their research on anaemia, Mol and Law (*ibid*) identify three, non-exclusive, non-exhaustive types of emplacement—that is, types of topologies: regional, network and fluid. Crucially, the different topologies are not mutually exclusive. They co-exist, and their particular constitutions are emergent from the relations between them (*ibid*: 663).

Regional topologies are created when the differences within a geographical region are subsumed to create a homogeneity that matches geographical, metric borders (*ibid*: 646). Thus, the organisational structure of FoEI based on national member groups, as well as their endorsement of geographically distinct ecosystems, resonates closely with 'regional topology'. In the office of FoE Brazil in Porto Alegre, Brazil, Mariangela had just downloaded Google Earth on to the computer where she worked. Amy, Mariangela and myself sat around the computer as they tried it out. They used it to identify the locations of their activities in Porto Alegre, in other parts of Brazil and to find my house in Malta. We spent the afternoon zooming in to different regions of Brazil and of the world. Mariangela particularly enjoyed the way the names of cities, towns and localities appeared as the satellite image on the screen zoomed closer to the ground, the way that when the image zoomed out to view the world from 'Space' you could click on any visible land mass and the 3-D globe image would rotate smoothly, simultaneously beginning to zoom in, the land taking up more and more space on the computer screen, until the names of countries, then cities, then towns began to appear. Google Earth is one of the visualising technologies that, according to Haraway (1991: 189), disseminate the illusion of a vision disembodied from situated positions. It is through Google Earth, the world maps on the walls of their office, aerial photography, plane travel and other such experiences that FoEI activists' experience is incorporated on a daily basis into a regional topology.

By contrast, in a network topology proximity is *not* metric. Rather, it depends on access to parts of a system (or an infrastructure) (Mol and Law 1994: 649). The FoEI activists' *close* ties, developed over the Internet, telephone or other communication technologies, exist in and create a network topology. During a FoEI general meeting in Nigeria in 2006, the Internet was only available on two computers within the area to which the activists had access. Most of the activists were not used to this. Whereas normally they would be following international news websites, the activists came to know of the news that North Korea had officially launched a nuclear programme through a text message (which was also only sporadically available) sent to the representative of FoE Japan. Someone jokingly commented that not having their usual access to communication media felt like they were having the meeting in a bubble. Calhoun (cited in Hannerz 1996: 248) argues that Internet and telephone are indirect means of interaction. However, many FoEI activists, indeed many people who regularly use the Internet and telephony (Rouse 2002), do not feel it is indirect. While they do not confuse it with face-to-face presence, the various communication media are experienced as offering distinct types of presence. This is a different experience of place; a network topology. However, it can only successfully occur in the way it is currently experienced by FoEI activists if, among other things, the landscape (for mobile phone reception), the tools (for Internet access to computers, plus climate control since the computers would not work for long in hot and humid conditions) and the infrastructure (access to a relatively stable electrical supply) co-operate. The meeting in Nigeria felt like an isolated 'bubble' because certain constituents of the environment did not afford the kind of presence co-created in a network topology. This topology is so central to the constitution and maintenance of FoEI that Chapter 7 is dedicated entirely to the kind of place and presence developed with communication technologies.

Finally, whereas in network topology what creates 'places' are sets of similar elements and similar relations between them, fluid topology denotes experiences where both the elements and the relations between them can vary, while remaining recognisably distinct. Fluid topology is defined by this 'invariant transformation' (Mol and Law 1994: 658). When in 2003 I met Anastasia, the coordinator of FoE Cyprus, in a project staff meeting in Malta, we were the only two FoE activists among a group of environmentalists from other groups and contexts. After having given my presentation about FoE Malta and our particular work in the project, Anastasia approached me and set about asking for news of other FoE Malta members whom she had gotten to know over the years. We had never met before, yet she spoke to me with familiarity and pointed out who other people were, and told me stories about FoEI meetings and about environmentalist politics. Although the proximity that I experienced could be taken as 'social', in the sense that what counted to Anastasia was the FoEI diacritic of identity, by attending subsequent meetings and meeting more FoEI activists there I realised that Anastasia was teaching me how to behave as a member of FoEI. However, I was not only learning with my social person, my whole self was learning; learning how to sit slightly inclined towards the other FoEI activist, learning how to look and notice if others had a FoEI sticker on their laptop or diaries, learning how to show welcome to fellow FoEI activists through an enthusiasm communicated not in words but in orientation—expressing interest with the whole body.

The different types of emplacement that I encountered with FoEI activists show that sociality needs to be understood as being mutually constituted within a field of relations in which the FoE activists, as whole organism-persons, are engaged, rather than established between disembodied social personae. Moreover, due to the different environments that activists constantly encounter, their experience of place, location or sites is far from isotropic (the same everywhere). Different types of presence or emplacement can be felt simultaneously, and these places change, depending on the environment the activists are relating to, including, but not exclusively, other people.

Conclusion

In the process of theorising about mobility and flow—as opposed to the implicit orthodox immobility and boundedness of the 'field'—multi-sited practice/theory has so far proposed a further detachment of sociality from emplaced lived experience. This solution perpetuates a theoretically obsolete division between mind and world. I have attempted to show how the 'multi-sited' notion that people's social life is detached from their environmental relations, now more than ever due to increased mobility, reproduces the same social theory implicit in the Malinowskian fieldwork ideal that couples modernity with such detachment. However, the debates that multi-sited practice/theory have stimulated have drawn attention to the issue of 'sited-ness', which for my research project provided a backdrop against which to explore how FoEI activists relate to their environments. Through my analysis of some aspects of FoEI activists' experiences, I have tried to show that it is necessary to further develop our sensitivity to how people relate to their experience of emplacement, and to the environmental relations together with which such places are co-produced. This shows that there is more to say about people's experience of 'place', at least that of FoEI activists, than a detachment of the social from the environment allows.

Mol and Law do not argue for a revolution of theoretical assumptions, replacing the 'new' fluid, for the old regions and networks. Rather, they argue for 'topological multiplicity' (1994: 644). Fluid, regional and network topologies are not essences—they are states in time. A state can change. Neither state is primary, or fundamental. What is fundamental is that the state of something is mutable. It may seem that this amounts to saying that one state, the fluid, which sometimes congeals into the regional and at others into the network, underlies and is common to both. This, however, is not the case. The fluid topology has specific characteristics, which are not the same as the potential for mutability in all the states. The definition of a fluid topology is the experience that 'both objects and relations are mutable'. Fluid topology creates 'invariant transformation', which means that the objects and their relations can change. However, they will continue to be considered parts of the same thing. Furthermore, different fluid topologies do not mix chaotically, or two states may coexist without merging. These different topologies are recognisable in the activists' experiences and ideologies of 'scale' and place.

In conclusion, force fields help to conceptualise the ongoing relations (with varying rates of change and stability) between the different topological states I observed at play during my fieldwork. Force fields do not imply the predominance of one or another state (fluid, regional or network, or others as yet not noted). The key is the mutability, influence *and* possible independence between states that the notion allows for. The implication is that it makes it possible to relate the experiences of the activists and the different states, without having to reduce any of the different states to one or other of them.

Notes

1 Following Anna Järpe's (2007) suggestion that photos are useful tools of documentation and are, in themselves, also field notes.
2 The term 'multi-sited' includes research that is defined as transnational, for this reason I use the term 'multi-sited practice/theory' to include the literature cited about transnational research.
3 For an exposition of how learning is an education of attention in a person's total field of relations, see Ingold (1993, 2001).
4 As an aside, although Coleman and Collins suggest that it is the loss of regional comparison in the synecdoche global/local that creates uncertainty within and about the boundaries of the field of anthropology, the 'global' has always been implicit in the anthropological comparative project. In its claim to transcend ethnocentricity, anthropology has often aimed at being 'global', with all the universalisms that entails, even more so while the cultures it studies are considered local (see Ingold 1993). Thus it is the blurring of the global/local distinction, for instance in terms such as 'glocalsation' (Robertson 1995), that destabilise the traditional comparative method of anthropology. I am indebted to Tim Ingold for making this point.

4 Chronological History and Organic Time

Being Introduced to FoEI

Volunteer Beginnings

The first thing a new volunteer of Núcleo Amigos da Terra (NAT or FoE Brazil) is given to do is to read the history of the organisation. On my first day at the NAT office, Andrea, coordinator of NAT, handed me a paper folder containing a dozen or so sheets of text, which on inspection had obviously been handled by many people. The papers were yellowing and bent at the corners where many fingers had turned the pages. It was clear that I was expected to read these histories. Andrea asked me two or three times during that first week whether I had done so.

As I walked into the front room of the NAT office for that first meeting, it was also impossible not to notice that the walls were lined floor to ceiling with metal bookshelves, heavy with catalogued books, magazines and pamphlets. After I had been handed the history of NAT, Emilia, another NAT activist, guided me to one of the bookshelves and pulled out two books for me to take home; both were histories of environmentalism in Brazil. At the time, I was not surprised by the NAT activists' emphasis on history as the first thing I was handed on my employment with FoE Malta in 2003 was a similar well-handled paper document of the history of FoE Malta. And, subsequently, I had handed out these histories to new volunteers with FoE Malta myself and passionately asked that they read them, just as Andrea then did with me and just as Judith did with any new volunteer with the FoEI International Secretariat. For FoE activists I worked with, knowing the history of the organisation, or the International Federation, is the first step to becoming a part of that organisation and, in time, an activist who will add to that history; adding to the campaigns, activities and documents that are understood to compose these histories.

Particular apprehensions of history and particular forms of 'tradition' are and have been central in the creation and establishment of authority and legitimacy that lead to very particular sorts of groupings. Anderson (2006) and Hobsbawm and Ranger (1983) argue that the logic of the nation-state depends on two concomitant notions of the past. First, the past of a nation stretches backward from the present in a way that ensures continuity but prevents return visits. The understanding of time is spatial, so that the past is behind us, but also cannot be accessed (Hodges 2008). Second, however, only those with reflexive, overarching knowledge of a nation's unitary history can proffer its value as 'National Patrimony' (Briggs 1996). According to Briggs, this has been the 'major force in the politics of culture' since Herder in the later eighteenth century (*ibid*). Embedded in the institutional workings of FoE groups and of FoEI, the way history is understood is closely tied with the nation-state, national consciousness, and the system of nation-states. The temporal way of thinking that builds national consciousness is directed towards building a sense of belonging to the national FoE organisations, as well as towards building legitimacy for the organisations in their respective countries. Those national member groups are then assembled in Friends of the Earth International.

However, FoE activists juggle different conceptualisations of time in their understandings, uses, experiences and negotiations of the past and of history. In this first ethnographic chapter, starting with 'history' to reflect the way new volunteers are introduced to the respective organisations, I explore the different experiences and concepts of time that FoE activists employ and relate in their documents and how this affects the form of the organisation. At times, FoE activists talk of their organisational histories in terms of monuments with founders and founding principles that remain unchanged and move through time as bounded entities, thus echoing the way history is conceived within national consciousness (Anderson 2006) and the environments of legitimation within which NGOs such as FoE are constituted as entities. Time in this framework is empty. Stable bodies such as nations and organisations 'travel' through it unaffected by time itself (Ingold and Hallam 2007; Hodges 2008). Following this logic change can only be effected by 'events' (Strathern 1990). More specifically, change and creativity are equated with innovation and understood as such only when creative acts break with history; to be creative (understood as innovation) is to introduce a discontinuity with the past (Ingold and Hallam 2007).

The result is an apparent double bind for FoE activists. On the one hand, the environments of legitimation in which FoE member groups function, namely their respective nation-states, equate a long history with reliability and stability (the two go hand in hand in such expectations). On the other hand, FoE activists would like their groups to be responsive, adaptive to changing circumstances and inclusive of the possibly new perspectives that new activists bring with them. Yet they do not want to lose the history of work that they were attracted to as new volunteers and the reputation that 'history' gives to an organisation. Furthermore, one of the points of agreement amongst the FoE activists I worked with was that continual growth and adaptation are inherent in how the world works. In addition, growth and adaptation are ideals to work towards.

In some circumstances, time and history are understood to be one and the same thing, as growth, where the past is not 'a foreign country'. When FoE activists carry on the work of others and their own previous work, those past relationships are what make the present and are present in tangible

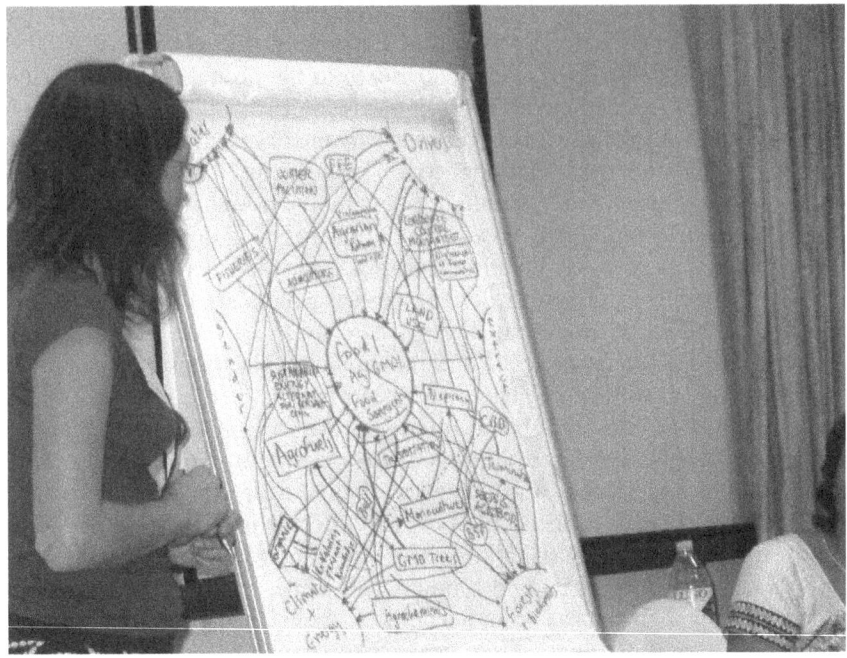

Figure 4.1 At the Swaziland FoEI inter-BGM 2007.

Note: A planning meeting to restructure FoEI's campaigns and programmes according to the new ten-year strategy developed through the SVPP. The chart shows the multiple links between FoEI campaigns. When this group presented their work to the plenary everyone laughed, and the presenter joked, 'Only environmentalists could end a planning meeting with the conclusion that everything is related to everything else'.

terms. Documents, books, posters, files and prizes, among other things, render visible and tangible other activists' (and their own) past work. That past and ongoing work is understood as what constitutes the organisation today. In this conception 'time' is not empty, it is not a separate dimension in which things happen. Change inheres in the relationships that people and things engage in that have 'duration' (Whitehead), and as such continuity and change are not opposed.

In this understanding, however, the boundaries of and within the organisation are not clear (see Figure 4.1). The notion of growth indicates neither a clear creation point of the organisation, nor exceptional founding events, nor a clear distinction from relationships that are not considered 'FoE work'. The notion of growth does not emphasise the creation of distinct entities: densities yes, but not bounded objects. In the history of civic action, scholars, governments and law courts have tended to concede legitimacy only to entities. Differing types of densities, rather than solid organisations, are not considered 'civic society' in the eyes of governments and some academic disciplines (Hann 1996).

Oriented Towards the Nation-State

To be considered for membership to FoEI, applicant groups must be already 'working on the main environmental issues in their countries' and 'working on both the national and the grassroots levels'.[1] FoEI membership allows only one FoE group per country. FoEI is a network whose strategy is to function simultaneously in national and international fora. For this simultaneity to work requires that member groups already have the attention of their national governments and publics. All the groups that are currently FoEI members were set up with reference to nation-states prior to becoming members of the international FoE Federation. In fact, only around half the current seventy-six groups use the name Friends of the Earth.[2] The member groups were founded to address the nation-states whose laws and actions they campaign against, or whose officials and politicians they either expose, cooperate with or try to lobby. However, this orientation towards their respective nation-states informs the constitution of the national organisations in more subtle ways than simply the reach and intended audience of their work. The organisational histories of FoE Malta, FoE Brazil and the IS mirror the time reckoning and historical forms that are also fundamental to national consciousness.

Hobsbawm (1983: 2) argues that the 'invention of tradition' has the effect of establishing continuity with a suitable historic past. He argues that for political groups and movements in the nineteenth century, national consciousness was among one of the changes that were so unprecedented that historic continuity had to be invented (*ibid*: 7). Despite the difficulties in Hobsbawm's arguments,[3] he shows how in the creation, establishment or maintenance of groups, continuity with the past is a powerful tool for authority and legitimacy (*ibid*: 9). In addition, continuity in the form of history is particularly associated with national consciousness. Terence Ranger (1993) later argued specifically for a link to be made between imagined communities and the invention of tradition.

Anderson (2006) argues that one of the key changes in consciousness that contributed to the emergence of national consciousness was a change in the conception of time (*ibid*: 22): from a messianic simultaneity characteristic of the older religious and dynastic realms, where past and future are in the same instance part of the present, to the notion of 'homogeneous, empty time', in which parallel imagined communities, nations, can move through history as solid entities (*ibid*: 24–26). Therefore, a particular type of apprehension of time and a corresponding form of historiography (Hirsch and Stewart 2005: 265) are tightly entangled with the rise of national consciousness and subsequently of nation-states.

National consciousness and the model of the nation-state informs the political struggles of some FoE activists and those of the people they work with. Central to the identity politics of indigenous peoples is the notion of a coherent and unique past, ascertained by scholars through archives and documents (not only narrated by elders).[4] These have been imposed through the spread of the system of nation-states and the rulings of Western courts of law (Hanson 1997; Briggs 1996). History and tradition are the primary tools in establishing the continuity with the past that within postcolonial courts and governments gives legitimacy to the claims of such groupings (Weiner 2007; Anderson and Berglund 2003; Hanson 1997; Briggs 1996; Campbell 1993; Linnekin 1992). Many FoE activists are not only familiar with these political groups but in many cases they are the 'grassroots' groups and 'social movements' with which FoE groups are allied. At FoEI international meetings, most of the issues to do with land, forest and marine exploitation include FoE groups that work with the people from whom the land, forests and seas were appropriated, or whose land is being polluted by adjacent extractive or manufacturing industries. The most iconic example is FoE Nigeria's work with Ogoni people in the Niger Delta, who frame their struggles against the multination petroleum industry and the Nigerian government in the terms of identity politics.[5]

The FoE histories are also enmeshed with national consciousness in the forms of historicism they take. Hirsch and Stewart (2005) describe the particular historicism, associated with the 'West', as the notion that there is a single 'real' past that is separated from the present. They argue that what was until the French revolution the particular struggle of the humanist political movement against the papacy and the Holy Roman Emperors, has become understood as a general quality of human existence (*ibid*: 263–4). Ranke was part of this movement. He warned against anachronism and described the past as 'a foreign country steeped in its particular institutions, culture and forms of thought' (cited in Hirsch and Stewart 2005: 264). In addition, change over time can be explained by historical development according to Rankean historicism; in other words, history is the accumulation of causes and effects. Ranke is often placed at the genesis of scholarly history, which subsequently becomes 'real history' (*ibid*) and opposed to what are called myths, legends or stories of oral history or non-orthodox historical endeavours (Wogan 2001; Vansina 1985). The histories of FoE Malta, FoE Brazil and the IS, as they are written, are mostly Rankean histories. The organisations are depicted as being the outcome of cumulative processes; their activities, campaigns, projects and so on, all the achievements of the past, understood as a real, objective and unchangeable past.

This does not mean that were it not for their engagement with national governments, FoE groups would have no use for archives, books, documents of their campaigns, and so on. The FoE Brazil volunteer and pensioner, Amy, who archives all mentions of the organisation in the newspapers, cultivates those cuttings for a variety of reasons. One reason is that Amy, as the curator of these cuttings, has a very clear role in the organisation especially since she feels the organisation has been, and is still, changing very fast, too fast for her. This reason is not directly related to the success of FoE Brazil as a 'national' NGO. The FoE Malta timeline was turned into an exhibition to be displayed at various public events. Yet it began as a means to reinforce and recreate group coherence when Julian Manduca, a previous coordinator of FoE Malta, died and the group lost the glue that, in their perception, held the organisation together. The orientation to the nation-state does mean, however, that specific community-building tools are readily available for pirating and adapting to the group's particular uses. It also explains the particular channels used for communicating to those outside the group and those new to it.

Temporal Markers

Different types of temporal markers help form the consistent 'historical' identity of FoE groups. The parallel aspects of national consciousness and FoE historicity that come across most strongly are: the emphasis on the origins of the organisation, a particular use of chronology, and the understanding of history as being cumulative.

Origins and Continuity

During their first week, new members of FoE Brazil are only given the history to read.[6] New members in Malta and the FoEI IS, however, are given current and recent publications as well as a history.[7] Currently, FoE Brazil is more concerned with history than FoE Malta and the IS. There are specific political reasons why history is de-emphasised in the IS.

The difference in the degree of emphasis on history also highlights the different contexts in which these organisations operate. In Brazil, especially in Rio Grande do Sul, environmental organisations claim to be pioneers of environmentalism. In this region producing documents, the activities of public intellectuals, ENGOs, organised protests and campaigns were all part of public life from the 1930s. Rio Grand do Sul, or Gaucho (as inhabitants from this region are called), environmentalism was also of considerable influence for the rest of Brazil (Urban 2001). José Lutzenberger,[8] the figurehead of this Gaucho movement, became minister in the newly created Ministry of the Environment in 1990. Environmentalist discourse in Rio Grande do Sul is imbued with the doctrine that current legitimacy is based on pioneer qualities. Their discourse is about being the first to think and act as organised environmentalists.

The founders of the original group that is today called NAT are also considered pioneers. In the NAT history texts, the founder of the original group, called ADFG (Ação Democrática Feminina Gaúcha), is Magda Renner (her real name). Together with three other women, Magda is described as working for the education and empowerment of underprivileged women of Porto Alegre. The early activities of ADFG included organizing evening classes for women and other educational events for disadvantaged women. The history texts also point to Magda as the source of the change from feminist issues to more environmental ones. Magda is depicted as having heard Lutzenberger speaking and being entirely convinced that issues of democracy and wellbeing cannot be advanced if multinational chemical companies pollute the food and water that housewives manage and offer to their families. The history text identifies this as the beginnings of eco-feminism that then spread to the rest of Brazil. When ADFG applied to join FoE International in 1985, the women were required to open membership and the focus of the organisation beyond women and feminine issues. This is when the name was changed to NAT.

The FoE Brazil history, as it is recounted by the activists and the documents they disseminate, depicts the organisation as no longer composed primarily of the wives of rich and powerful men. However, the women's past work counts as part of the same organisation. FoE Malta, the IS and FoE Brazil converge in sourcing their organisations to founding personas. All three groupings begin their histories with a point of origin. Here are the first lines from the history documents of FoEI, FoE Malta and FoE Brazil:

- Friends of the Earth International (FoEI) was founded in 1971 by four organisations from France, Sweden, England and the United States.
- Friends of the Earth Malta has been active in the environmental field since 1985 under the name of *Żghażagħ għall-Ambjent* (Youth for the Environment) and later as *Moviment għall-Ambjent* (Movement for the Environment).
- The Nucleus Friends of the Earth Brazil (NAT/Brazil) is an OSCIP—an official category of organisation charged with care for public interest—based in Porto Alegre, in Rio Grande do Sul, that has been active in the protection of the environment for more than forty years.[9]

The emphasis on a specific moment of creation within living memory is different from the way nationalism extends membership to past, present and future—as though the nation had always existed, emerging from the mists of time (Anderson 2006). However, what is similar is the use of history by an organisation to define itself. Anderson (*ibid*: 146) makes the point that national consciousness is simultaneously open and closed: closed as when, for instance, a particular language cuts off who can be included, but open because a language is always inviting new people to enter into conversation through it. The FoE histories of their founders establish a point of origin in the past that creates a distinct identity in a similar logic of identities within the consciousness of the nation-state (as discontinuous and bounded). A retrospective reference to founders' values or exceptional events in the organisation's history is part of boundary-making (Batteau 2001: 279). But at the same time, a founding story is open because history is understood to accumulate as it goes on. New activists are invited to join and carry the organisation forward. The uninterrupted activities in FoE history documents reiterate what is implicit in having a history, that the 'organisation' has continuity and coherence over time. New activists may bring new directions and new forms of work, for instance shifting NAT from a volunteer and member-based organisation to a project-oriented professional organisation, or shifting FoE Malta from a multi-issue lobby group to a more or less single campaign organisation. However, 'the' organisation is understood to remain intact.

The openness that allows for change without losing the distinctiveness of the organisation is described with metaphors of growth. Here are three more extracts from the three FoE groups that show this notion of open coherence (in each, the italics are added):

- FoEI: '*Today's Federation* of seventy-seven groups *grew* from annual meetings of environmentalists from different countries who agreed to campaign together on certain crucial issues, such as nuclear energy and whaling.'
- NAT Brasil: 'With forty years of uninterrupted activities, it *is one of the oldest* environmental organisations in Brazil. It was founded in 1964, with the name Ação Democrática Feminina Gaúcha (ADFG), dedicated to the promotion of good citizenship, through social and educational programmes, initially directed at women, primarily in lower income brackets.'[10]
- FoE Malta: 'Moviment għall-Ambjent, Friends of the Earth (Malta), formerly Żgħażagħ għall-Ambjent, *has been active* in the environmental field since 1985.'

Before I expand on the notion of growth let me explore how chronology is used, since the different notions are presented seamlessly in the same FoE documents.

Chronology and Events

Chronological order describes continuity as a line of significant events judged as such retrospectively, from the present looking back to the past. In the historical texts of all three FoE groups' activities, projects, campaigns or events are related in chronological order. The activities described are considered exemplary, embodying the essence of the organisation. In the texts, 'events' are understood in the Rankean sense of defining moments and actions that 'add up' to a history when analysed in retrospect. In FoEI's *First 25 Years* publication and in NAT's history document, events are described within divisions of decades, starting with the 1960s for NAT, and the 1970s, for FoEI, going on through the 1980s and the 1990s. Both FoE Malta history documents describe activities year by year. The older version was written by Julian Manduca, who was also one of the founders of Żgħażagħ għall-Ambjent (Youth for the Environment). Then the text was updated every so often, and the newer version was made by myself and two other more recent activists in 2005. We followed the chronology of the older history (sentences were cut and pasted from it) and deliberately mimicked the history activity held at the FoEI BGM in Malaysia 2005 as Anthony had described it. For the twentieth anniversary history exhibition text, we used separate sheets of paper, each representing a year of FoE Malta's existence. A green line curls from page to page, depicting a stylised bean sprout, which was used to symbolise the year-on-year growth of the organisation. Incidentally, when I had made that design, deliberately employing the metaphor of organic growth, I was unaware of the impression, conveyed by the analogy, that the people who composed FoE constituted a single organism. Nor was I aware of the implications for apprehensions of time, in which the differences between chronology and growth were being elided. The posters mingled chronology and growth as though they were undoubtedly the same thing.

The campaigns are described in the third person, as pertaining to 'the organisation'. Only Julian Manduca is mentioned by name, mainly because he had just passed away. The name of the organisation works in the same way as the symbol of the nation—the flag (Anderson 2006). The symbol 'FoE Malta' is inclusive because it is ambiguous, even if the activists know who organised those activities, sweated to carry those banners or costumes, spent hours researching, talking to people, thinking about issues, writing letters and discussing issues with people to collect signatures for petitions. Again, like the nation, 'the organisation' is simultaneously open and closed (*ibid*: 146). No names are mentioned, so it is closed, opaque, a thing. Yet in this way it is open and allows more people to join and identify with 'the' organisation.

In the introductory video on FoE Malta's website,[11] activities that in previous documents were represented according to the year they happened, were grouped together under themes and campaigns. This video is now the section that gives information about FoE Malta's history. Here the 'who we are' section is merged with the old history text. There is no separate section for history, as there used to be on previous FoE Malta websites and as there is on FoEI and NAT websites. A similar mixing of 'history' and 'about us' sections can be found in the 2001 FoEI document celebrating thirty years. Therefore, chronology is not always used in boundary-making activities.

In the NAT history text on the website, which is also given to new volunteers, the chronology is often mixed up. The text moves backward and forward in time, where quotes from interviews held at one date refer to events prior to the interview. In these instances the back and forth movement is clearly marked with the dates of the interview. The dates highlight an attention to chronology. Chronology is important to show continuity to others or newcomers, whereas experienced activists feel the continuity and are not so dependent on its chronological depiction.

Cumulative History and Growth

The difference between the expression of cumulative history, respectively, by chronology and by growth, is that the former is an accounting of events *in retrospect*, whereas the latter *carries on* with moments of punctuation

(Ingold 2007a). There are intense moments in the work and daily lives of FoE activists with protests, festivals, meetings, project endings or beginnings. There are also quieter periods, seeing to the daily accounts, catching up on emails, reading and discussing in the office or going for walks in the countryside. In the metaphor of growth, there are no complete breaks between these differently textured goings on.

When the FoE activists I worked with spoke about their respective organisations, there was no clear disjuncture between the present and the past. At times, the 'direction' of the organisation would be changing. For instance, after Julian Manduca's untimely death, Anthony proposed that FoE Malta join FoEE's capacity-building project. In turn, this meant retreating largely from public activity to develop its capacity as well as recruiting more volunteers, employing a 'change agent' and routinising a lot of the work. The long-standing members were explicit about how FoE Malta was changing. Land use was one of the issues that FoE Malta focused on for many years. In 2006, a number of FoE Malta activists, mainly long-standing, complained to Anthony that the group seemed to be losing the focus on land issues. With the capacity-building project, 'FoE Malta' was being asked to single out one area. Rather than dedicating energy to the six or seven other campaigns that were previously running simultaneously, FoE Malta was to choose a single focus. In 2007, the activists in Malta agreed to focus on climate change. Other issues such as land use would be included where they were relevant to climate change. And yet, even with all these changes there was no doubt in the long-standing activists' minds, who were critical of these changes, that this was 'the same' organisation.

In FoE Brazil, a similar change in the nature of the organisation came with the shift from a volunteer-based to a project-based organisation. Another, even more radical change, was the shift from a feminist to an environmental organisation in the mid-1980s. With this latter shift, even the name of the organisation was changed. Yet, the organisation is understood and experienced to be the same. The written texts help solidify this continuity for outsiders. Consider for instance: *Núcleo Amigos da Terra/Brasil . . . was founded in 1964, with the name Ação Democrática Feminina Gaúcha (ADFG)*'.[12] Here, it was NAT that was founded, the different name does not change this in any way. It is similar with FoE Malta: '*Moviment għall-Ambjent*, Friends of the Earth (Malta), formerly *Żgħażagħ għall-Ambjent*'. In an interview, one of the early activists in the group said that the word 'youth' was changed to 'movement' in the name of the organisation because the activists who had started as university students could no longer be called youths anymore. Here, we find again the notion of growth and of the organisation growing as an entity. As a person grows from youth to adult, the organisation matures *pari passu*, and yet survives each activist.

The organisations are understood as growing in this sense, that what each organisation is today is all the activities that were done in its name in the past; however this growth is not necessarily evolutionary. There are moments or situations when there is a sense that there has been an improvement compared with how things were done in the past. However, the respective directions of the particular FoE member groups are contested. Emilia thinks it is better that FoE Brazil now has paid employees, because this allows people from different economic backgrounds to have a voice in the environmental movement. Others yearn for the solidarity they experienced when the group was based on volunteers and had a strong membership base. A number of FoE Malta activists consider the narrowing of focus a necessary evil, not a straightforward improvement.

Pirate Origin

Even though the histories embody the continuity of the FoE groups that activists feel they belong to and continually constitute, the history texts are also strategic, ready-to-hand tools. We have already seen an example of how positioning the advisory committee members as part of the history of NAT increases their separation from the daily workings of NAT. We have also seen how removing chronology from FoE Malta's 'who we are' video bumps up the perceived activity of the organisation and as will be discussed in Chapter 3 'being active' is morally loaded in the environments of legitimation of NGOs. Furthermore, the founding of the International Federation is presented as strategically adopting, or pirating in Anderson's (2006) terms, the form of the system of nation-states to promote very different goals.

> [T]he 'nation' proved an invention on which it was impossible to secure a patent. It became widely available for pirating by widely different, and sometimes unexpected, hands.
>
> Anderson (2006: 67)

In this passage Anderson refers to the spread of the model of the nation-state from the Americas to Europe and then to other parts of the world. It is not simply by means of the narrative form of their official history documents that the FoE activists effect the orientation of their organisations towards the nation-state. I have already pointed to the symbolic and other in which ways retrospective historical accounts, chronology and founders are deployed in FoE documents. Beyond this, the activists' discussions, policies and history texts often make explicit links between particular nation-states or international institutions and the origins, aims and development of their respective FoE organisations.

David Brower, the founder of Friends of the Earth International, has reiterated in various interviews carried out by FoE activists over the years that FoEI was set up to monitor the work of international bodies, at the time, especially the International Maritime Organisation and the Whaling Committee within it, the United Nations' Economic Commission for Europe (ECE) Geneva Convention on Long-Range Transboundary Air Pollution (ratified in 1979), and others such as the UN General Assembly and meetings such as the 1972 Stockholm conference on the Human Environment (from Link Magazine, FoEI Archives). Later, the focus of FoEI spread to the International Monetary Fund (IMF), the World Bank (WB), while regional FoE affiliated groups are watchdogs of the EU (such as Bankwatch based in the Czech Republic), the Asia Development Bank, the South American Development Bank and the Initiative for the Integration of the Regional Infrastructure (South America) (IIRSA).[13] The FoEI Federation was originally set up as a watchdog for such international fora. The original purpose has not been dropped. It is considered to have developed and strengthened as the number of members of the Federation increased. The FoEI Federation still holds in common the conviction that to fulfil the aims of the Federation 'require[s] both strong grassroots activism and effective national and international campaigning and coordination' (FoEI history document *First 25 Years*, and the FoE Malta history until 2005). However, in addition to mirroring and following such international institutions, the early aim was to model their Federation on 'international' cooperation and to improve upon it.

The following is a passage from the history page on a past version of FoEI's website (italics added).

In 1969, the Executive Director of the US Sierra Club resigned out of frustration that the organisation neglected to tackle nuclear issues, or even to work internationally. This visionary man was called David Brower, and he explained:

Realising it was time to stop working toward a moon-like earth, I started a new organisation. We fished around for a name, and came up with Friends of the Earth. It was essential that it be international in scope. With meetings in London, Paris and Stockholm, we were able to convince environmental people in three more countries to let the FoE idea migrate. Other countries now in the FoE network were courted, or courted us, and in no time the sun was rising somewhere on a FoE group. *We made it a point not to be clearly organised or directed by some old tired formula from the top. Find good people with the right ideas and let them move ahead their way.*

(FoEI Archives Amsterdam, 1995)

The following passage from the same FoEI history document shows how the idea of improving on the model of international cooperation was carried forward by others even when David Brower was no longer involved with FoEI (italics added).

Ten years ago, I went to my first meeting of Friends of the Earth International. I recall a babble of accents, a kaleidoscope of ideas, views and strategies. The spectrum of resources at our command ranged from modest to tiny. Could this small, motley crew help save the Earth? But

I also remember the words of a Japanese member tumbling out so fast we had to ask her to slow down. She was pleading passionately, not for an issue in her own country, but for Pacific islanders threatened by nuclear testing. Not just their crisis: ours, too. Citizens' groups such as FoEI can reach across geographical and cultural boundaries, *to act together in a way that our governments have so often failed to do.*

First FoEI Chair Mairi MacArthur in *Save the Earth*
by Jonathon Porrit, 1991.

Even if FoEI is an international federation, the implicit strength of national consciousness requires me to emphasise the continued importance of the nation-state. Especially since, in many theories of globalisation, International NGOs (INGOs) are often seen as marking the beginning of the downfall of the system of nation-states (Castells 1997). Castells (*ibid*) predicts that in what he describes as an increasingly globalised world INGOs will replace nation states as the providers of wellbeing; replacements for the welfare state.[14] However, the international is fundamental to national consciousness. The plurality of independent states is essential to the validity and generalisability of the model of the 'nation' (Anderson 2006: 81).

While new members to the national groups and the International Secretariat are introduced via organisational history, first timers at international meetings are not given any similar overall history of FoEI, neither in written nor in spoken form. In fact, references to the past during introductory moments of international meetings are only to recent happenings at similar international meetings. Apart from the fact that it is assumed that they already are FoE Members even if they are not necessarily familiar with FoEI, the lack of emphasis on history at international FoE meetings is due to lack of agreement about history; claiming to know what such a history is could cause disagreement among the members.

Strategic Histories

History texts are central to FoE activists' attempts to garner different forms of support from national and international institutions, as well as from other audiences. Yet the FoE activists do not mirror the discourses of their legitimising audiences passively. As with other ENGOs, FoE groups deploy particular discourses to increase and maintain their legitimacy in the eyes of their 'core publics' (Lister 2003; Choy 2005; Berglund 2001; Gatt 2001), the institutions of the respective nation-states and international institutions that FoE Malta, NAT and the IS of FoEI address. The where and when of the focus on history is a clue to understanding that their histories are strategic. The different FoE groups emphasise their histories with new volunteers, in parts of their website and in their annual reports to funders. Their histories are one of the tools they use to communicate their organisation to institutions and an imagined 'general public'. The various institutions they address within the context of their respective nation-states comprise the main audience for this form of communication from which they gain legitimacy.

History, and more specifically 'FoEI' history, is not a central concern at General Meetings nor of the Executive Committee (ExCom). For instance, as I mentioned, a core member of the IS has been trying to get a 'history project' to produce a booklet or a website similar to the *Early Years* booklet, for the past four years. She has found it difficult to find support from volunteers; it has never been raised as an issue at BGMs and the ExCom have not yet approved funding for it. There are many reasons for this, including the view that the FoEI Federation is a forward-looking network and the activists prefer to use the little time at international meetings to work on current or future campaigns rather than on analysing past ones. However, one of the central reasons is that it is clearly understood in the federation, especially by many of the representatives of South and Central American FoE groups who are scholars and public intellectuals, that history and the particular way of recording and analysing history, the emphasis on texts and archives, and the side-lining of non-Euro American history and of oral history, are closely tied up with Euro-American realms of power and with (neo)colonialism.

The various FoE histories are strategically deployed in the register of their core public to strengthen their legitimacy. However, just what constitutes the core public of FoEI is a highly contested issue. Whereas in the 1970s national governments and international meetings or institutions were the target of FoEI actions and campaigns, today there is increasing dissatisfaction with international bodies, especially the UN. There have long been discussions within the FoEI programmes targeting international financial institutions such as the World Trade Organisation (WTO) and the World Bank, about whether directing campaigns explicitly at these institutions implicitly validates their existence. Following the unsatisfactory outcome of the Kyoto process, the FoEI Climate Programme took a specific policy decision in 2007 not to focus on UN processes and instead to work more on 'community actions'.

In European FoE groups, the issue is also discussed, often informed by interaction with South American groups. For instance, at the AGM in Estonia in 2006, there was a long discussion which started when the plenary was trying to word a sentence—it started as 'local versus global'. A discussion followed about how local communities are often the cause of pollution or environmental degradation for other communities. Small community mines upstream from other communities were mentioned as an example.[15] At this plenary meeting, national governments and international regulatory bodies were considered equally important to community or local empowerment. In the end, the sentence was drafted to read 'the local in relation to the global'.[16] Many activists consider working with their respective nation-states the primary duty of FoE member groups.

The discussions among FoE activists about which institutions and which structures of government are best for environmental justice do not directly mention whether they should or should not use forms of historical narrative because of their involvement with the particular structures of power and government, or neo-colonialism. However, together with the shift of focus from UN meetings to community support, there has also been a shift in the types of documents created. Rather than lists of achievements on the front pages of FoEI webpages, 'community voices' are increasingly taking the forefront. There is a subtle shift in FoEI from describing the identity of the Federation in terms of past actions and campaigns and therefore textual, chronological histories, to being the channel for the voices of marginalised 'communities'. Videos, recordings and narratives of peoples' experiences do not therefore necessarily take the form of retrospective Rankean histories. The debate itself and the change that is occurring show how core publics, and the tools to engage them, specifically ways of recording what FoEI is, are ready-to-hand instruments.

The way the history texts of the national FoE groups are used in the introduction of new activists also highlights the particular strategies that seem to be widespread among the FoE activists I worked with. 'Culture' in organisational studies is often understood as the property of an organisation (Wright 1994; Jimenez 2007). The notion of property portrays an organisation as having 'a culture', rather than the way things are done and thought about in and through organisations being regarded as cultural in themselves. Organisational studies have from inception had a close relationship with the thinking of practising managers (Wright 1994: 3). Wright (*ibid*) points to the ubiquitous mission statement to show how this essentialised understanding of culture has wide purchase in the governmentality of organisation (Shore and Wright 1997). 'Having a culture' is not only descriptive, it is also prescriptive. In the light of my ethnographic material I extend this to 'having a history text'.

Foundations and Growth

In FoE groups, two notions of history and tradition are present. There is the root metaphor of history as something fixed and unchangeable in the past, where tradition implies continuity and coherence. This is evident in the emphasis on founders and founding principles, as well as in the use of chronology. Concurrently, there is the notion of a world in formation, with FoEI constantly changing and adapting not only to the changing needs of the environments the activists work for, but also adapting as new activists bring new ideas, new experiences, and in this openness allowing for strategic action.

On the one hand, the discourse on foundations and founders allows for distinctiveness, but restricts change. This is an understanding mainly imposed by law courts, funding bodies and political bodies towards which FoE groups and FoEI aim their work (see Weiner 2007; Mauzé 1997; Briggs 1996). On the other hand, the notion of growth allows for change and continuity but loses the distinctiveness and the boundedness of a monument-like organisation, fixed in stone.

The metaphor of foundations in FoE groupings is mainly deployed in dialogue with environments of legitimation and in recruiting new activists. The metaphor is deployed in situations where FoE groups need to be distinct for funding reasons or as an independent voice in consultation processes. The notion of growth, on the other hand, is more in line with most FoE activists' environmentalist ideology and its political corollary of inclusiveness. The notion of growth also allows FoEI to be more responsive than other International non-Governmental Organisations (INGOs) (Timmer 2007).

As I have described previously, at FoE International meetings there is less of an emphasis on history, and more emphasis on the present. The notion of international itself cuts across the temporal trajectories of the histories of nations, and requires a bracketing of history to be sustained. The paradox here is that while the logic of nationhood, and by extension FoE national member groups, depends on historical origins, the International Federation that these groups form in FoEI needs to partly obviate those national identities in order to function as a collective. The national identities of FoE activists, based on shared history, shared language and shared geopolitical states, are exclusive—in Anderson's (2006) terms they are closed. At FoEI meetings, activists very clearly do not share the same history, nationality and language, nor even similar approaches to environmentalism. However, in communicating FoEI to the general public, the IS and the ExCom does not have such concerns to create a sense of belonging. All these activists need to do is portray FoEI as coherent. In these portrayals, including in the history documents produced by the IS and ExCom so far, FoEI is coherent because it has a history.

The notion of a shared history was deployed by the IS and the ExCom when the federation was facing possible dissolution in 2003. The first task of the SVPP was the history task. Even the forward-looking, anticipating mission and vision partly produced through that task implies the past (Murphy 2010); the past embodied in memories and ongoing experiences of the injustices that the mission aims to redress.

Experience of Growth

In the international meetings and daily work of the activists, history in a Rankean sense is downplayed. The metaphor and experience of growth, on the other hand—on the past as present, swelling forwards—is ubiquitous in the offices, books and rhythmic routines. The activists see their actions accumulating and, over time, building the history of the organisation. This can be discerned in the layout of their offices and in the importance afforded to archives, posters of past events and other effects of past achievements.

The front room of the NAT office, lined with books, spoke volumes of the history that for new members is the first step into the organisation. In this way, new volunteers literally take their first step into the organisation through NAT's history, through the books, texts and accumulated knowledge. Anderson (2006: 44) argues that print contributes to giving a fixity central to the nation's image of antiquity. I would adapt his argument to say that the books, the documents and other things that you can find in NAT's front room are the tangible traces of the past. This is a past that inheres in the knowledge, offices, things, personal relationships and the organisation itself; the constitution of these things are the relationships and engagements of previous activists, and the previous work of current activists with their environments.

As the activists do their research, campaigns and projects they participate in various relationships: these are experience, knowledge and expertise in the making. That knowledge-making process includes these books and is subsequently turned into more books and leaflets. The full bookshelves embody part of FoE Brazil's knowledge, a material record of the years of research, activity and existence of the organisation. The front room of the NAT office is therefore a testament to its expertise; outsiders walk into a domain of knowledge and evidence of the organisation's existence over the years in which the books were collected. All this they do before the figurative first step they take when they trace the path of the organisation over the years by reading the histories. The history and the knowledge they step into in the office are communicated largely non-verbally; the office speaks a language of symbolism and of half-hinted impressions of authority and knowledge.

David Zammit (1998) shows how lawyers in Malta reflect the ideals they have of themselves and their professions in the way they organise their offices. The layout of the furniture, the position of books and clients' files and other objects in their offices embody the lawyers' professional detachment from clients (*ibid*, 145). This ideal is also embodied in the way the lawyers and their clients move through and sit in these offices. Similarly, the way visitors or new members are ushered through the FoE offices sheds light on how continuity is communicated and experienced non-verbally.

Unlike the Maltese lawyers, FoE groups do not have clients, nor do the activists hold professional detachment as their ideal. Their ideals as environmentalists are complex and contingent and I explore this in more detail in the next chapter. However, new members are presented with the ideal of an organisation that has a distinct history. It has an origin and is therefore distinct from other organisations. As the word 'organisation' suggests, this history can show how it is part of the organised politics of civil society, as opposed to amorphous non-organised social movements (Hann and Dunn 1996). The actions and activities of its members accumulate to become in an accrued way the identity of the organisation, which then *is* those values and actions in the present—chronology is dropped. In this way the new member, can become part of the organisation and in due course gain almost a sense of immortality that they will live on with the organisation even if not named in the actions listed. They may be even named, and become 'environmental saints' (Gatt 2001), as happened with founders like Magda Renner or key figures such as Julian Manduca. Although the organisations I worked with did not have the resources to plan their offices according to their ideal. Many spaces have to serve a number of different functions simultaneously. The front room of the NAT office is also where events for the paying members or members of the public were held such as the weekly lectures and discussions for NAT members called *Quarta Temáticas* (Wednesday Themes). It is also the room where their meetings with other NGOs are held, as well as the weekly staff meeting—because it is the only room where everyone fits and where the Conselho Diretor (the Director's Council) is convened once a month. However, the ways new members *are ushered* through the office, and to which points in the office, speaks of their ideal office layout. The plans and photos of the NAT/FoE Brazil office (see Figure 4.2) can be used in conjunction with this text to follow the new volunteers in their introduction to the organisation and get a sense of the non-verbal communication involved.

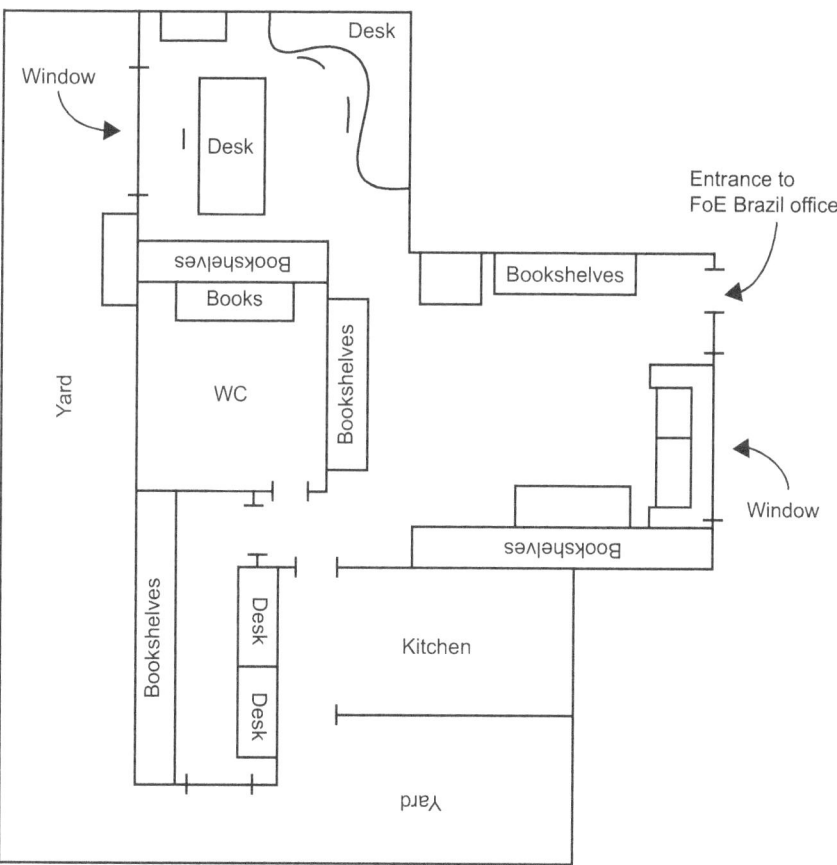

Figure 4.2 Plan of NAT/FoE Brazil office.

Apart from being the place where new volunteers wait to meet Andrea, the coordinator of FoE Brazil, it is also where they usually spend their first hours talking to Veronica and Amy, two long-standing activists, who make it a point to get to know new volunteers in that room—where they always sit and work and talk to each other and new volunteers. I spent my first week or so sitting in that front room, until specific tasks were assigned to me. The same happened with Jennifer, a new volunteer from Australia who joined a few months after me and became my flatmate. Next, a new member is shown the two inner rooms, the bathroom and the kitchen. The two inner rooms were radically different from the front room and from each other.

A large wooden desk with a green leather top dominated the main inner room and it stood in stark contrast to the other tables in the office mainly because it had no computer on top of it, only a telephone and some papers currently in use. It sat virtually in the middle of the room, with a large window and considerable space behind it. The L-shaped table lining the opposite wall held two computers a printer and a fax machine. In this room there were no high bookshelves. Instead, NAT posters and maps of Brazil hung on the walls. The other inner room was long and had a window at the far end on the short wall. This room appeared longer and narrower on account of the ceiling-high metal shelves running down one whole length. The shelves carried box after cardboard box, each identical to the other except for what was written on its label, and they were a little dusty. These were the archives of NAT. Box upon box of newspaper cuttings of mentions of NAT, documents produced or collected by NAT in the past, photos taken by NAT as part of their watchdog activity of parks and rivers projects and past accounts, among other things—this was Amy's domain. New volunteers were never taken in there first, and to my knowledge people who were neither members, staff nor volunteers were rarely invited in there. However, the boxes and shelves were visible from the front room through the door of the room which was never closed, except on the rare occasions. The subtlety of these dusty archives, just visible and clearly private, added a sense of wonder to the otherwise publicly displayed knowledge—it suggested secret wisdom.

NAT has engaged in a three-year project with a team of architects to develop a communally designed office closer to the centre of Porto Alegre, the Brazilian city in which they are based. In this office project, one of the main features will be a library, including a public section with books and paper materials, a digital archive offering public access to historical documents (newspaper cuttings, documents, letters, research papers, photos) and a separate archive in which the actual documents will be 'preserved for posterity'. The same public and private store of books and documents is planned for the new office design.

FoE Malta's Anthony has kept every single FoE-related email since 1999 on a separate hard drive that he upgrades every couple of years. The tangible and visible traces of FoE Malta's growth are mainly Anthony's, since he has been the key person carrying the continuity of FoE Malta through various changes in recent years, the biggest change of which was Julian's unexpected and untimely death. Of course, all the other long-standing activists, such as Raffaella and Peter, also carry in their persons and the things they have kept the relationships that constitute FoE Malta.[17] However, the fact that they are not as active currently and have taken those traces of collective FoE Malta growth with them, point to how the boundaries-making efforts of activists to define the organisation constantly leak away. Sometimes they come back.

Figure 4.3 FoE Malta during a protest in 2006.

Raffaella intermittently participates and brings those memories and experiences with her. At other times, the leaking away is tempered by attempts to pass experience to new activists through 'hand over' periods. Boundary-making attempts are not always focused on. In the photos below there is a clear distinction between meetings of different ENGOs when boundaries are not erected in comparison to the photo of FoE Malta during a protest when the entity is clear marked (see Figure 4.3).

Conclusion

The discussion around Hobsbawm and Ranger's 'invention of tradition' and Anderson's 'imagined communities' in the 1980s and 1990s raised questions about histories as comprising current interpretations rather than objective, verifiable facts. The notion of invention of tradition made space for agency and did away with the idea of historical determinism (Mauzé 1997; Hanson 1997; Briggs 1996). At the same time, however, the notion that modernity embodied dynamism and change because it broke with the conservatism of tradition, a notion still present in the work of Hobsbawm and Ranger (1983), reinforced the idea that change is opposed to continuity. In their work, and in that of other scholars who consider that some traditions are genuine while others are invented (Hanson 1997), the notions of creativity and change, and by extension strategic action, are associated with a break from tradition, history and the past (Ingold and Hallam 2007).

In the 1990s the processual paradigm that was developing in anthropology came to understand tradition as inherently dynamic (Armin Geertz 1997). Invention was inherent in all culture and tradition; 'tradition and culture are constantly in the process of renegotiation and redefinition, such that invention is a normal and inevitable part of the perpetuation and use of all culture and tradition' (Hanson 1997). Yet Hanson (*ibid*), and others (Linnekin 1992; Weiner 2007), found that their academic work along these lines was rejected by indigenous communities in New Zealand. It was rejected not only by the indigenous activists and scholars Hanson worked with, but also by the anthropology department at the University of Auckland because it threatened to dismantle the carefully constructed land claims, based on tradition, that had only recently begun to make some headway. Hanson (1997) suggests possible ways of avoiding this. One way, would be to avoid using a word such as 'invention', replacing it for instance with 'reformulation' (of tradition) (*ibid*: 202). That however, he goes on, would be to 'show the imperial hand' by assuming that indigenous scholars and activists cannot handle threatening analyses (*ibid*). He concludes by calling for academic work to be dialectical, such that ideas like invention of tradition could be included in conversations with the indigenous groups who found his earlier work threatening.

However, Briggs (1996) raises an important difficulty. If, in the environments within which both these indigenous groups, and FoE groups work, notions of invention and creativity are still understood as opposed to tradition, any suggestions of invention will be judged by those in power as 'fabricated'. The persistence of the logic of essential identities and Rankean-type history in the environments of legitimation in which FoE works makes Briggs's problematic all the more relevant to FoE activists.

In communication with the bodies that fund FoE groups, or the governments that invite FoE activists as consultants, or the public hearings that accept FoE activists as representatives on environmental issues, including at times the people affected, the hegemonic logic makes it difficult for FoE activists to present their organisations as changeful and relational without rupturing the coherence or the distinctiveness of the organisation. However, by seamlessly mixing chronology with notions of growth they manage to convey the two apparently contradictory notions. This allows them the flexibility they need to react to changing situations and the consistent continuity expected by their environments of legitimation.

Grove-White (1993: 25) has noted that NGOs often successfully use reductionist language to manipulate hegemonic discourses to their advantage. But this alone is not enough, for two reasons. First, the activists experience their ongoing activities as mutually growing themselves and the organisation they constitute. Tangible things such as books, posters, offices, and the routinised pathways through the offices for new volunteers, periodically bring this growth into focus. Even when not in focus, this growth is the ground from which activists develop their work. Secondly, the essentialised and detaching understanding of Rankean history does not resonate with the prevalent environmentalist ideology. In this ideology everything is related and continually in relation. The extension of this ideology to politics engenders notions of incorporating difference and change in the way power relations are organised within FoE groups and FoEI.

FoE groups handle 'the double-edged sword of invented traditions' (Mauzé 1997: 7) by mixing their metaphors. Apparent contradictions do not surface unless analysed for FoE activists because they touch on things that they experience and for which one metaphor would not be enough. Mosse and Lewis (2006: 5) argue that these sort of gaps between policy and practice are places of opportunity. In these gaps, NGO activists can respond to contingent situations, with competing logics that would not be acceptable to the rationalizing logic of policy. I believe that the mixing of metaphors in FoE activists' self-definitions take this a step further. By managing to slip in notions of growth, without losing their legitimacy within the wider hegemony, FoE activists begin to make space for alternative logics of time and history, and consequently different ways of establishing authority and legitimacy.

Notes

1 See www.foei.org/en/who-we-are/about/membercriteria.html last accessed 12th January 2017.
2 See www.foei.org/en/who-we-are/member-directory/groups-by-region accessed 12th January 2017.

3 Especially (1983: 5) where he opposes custom with invented tradition on the grounds that the industrialised world changes faster than so-called 'traditional societies'. See Rogister and Vergati's (2004: 202) arguments that the invention of tradition needs to be explored in all societies and at all times—not only in post-industrial revolution Europe. Also see Simpson (2004), Hanson (1997) and Briggs (1996), for other critiques of Hobsbawm and Ranger's notion of the 'invention of tradition'.

4 For an exception, see Kendrick (2009), where instead, the appeal in certain identity politics is based not on the claim of a distinct identity, but on the right to oppose domination as fundamental to all humans and therefore open to all.

5 Since the 1950s, the Ogoni, and other peoples living in the Niger Delta have faced severe pollution and land degradation caused by constant oil spills from pipes traversing their agricultural land and rivers and from gas flaring. Their efforts to protect their land and livelihoods as well as health and wellbeing have been persistently treated as criminal by the Nigerian government.

6 See http://natbrasil.org.br/historico.htm accessed 12th January 2007.

7 See www.foemalta.org/about.html accessed August 31st 2005. Webpage accessible today through http://web.archive.org

8 His actual name.

9 The original reads: 'O Núcleo Amigos da Terra Brasil (NAT/Brasil) é uma Organização da Sociedade Civil de Interesse Público (OSCIP) com sede em Porto Alegre, no Rio Grande do Sul, que atua há mais de 40 anos na defesa do meio ambiente.'

10 The original reads: 'Com quarenta anos de atividades ininterruptas, constitui uma das mais antigas organizações ambientalistas do Brasil. Foi fundada em 1964, com o nome de Ação Democrática Feminina Gaúcha (ADFG), dedicando-se à promoção da cidadania, através de programas sociais e educacionais, inicialmente dirigidos a mulheres, prioritariamente àquelas de baixa renda.'

11 www.foemalta.org/home/index.php/who-we-are accessed 15th August 2009.

12 The text in Portuguese reads: 'O *Núcleo Amigos da Terra/Brasil* . . . *[f]oi fundada em 1964, com o nome de Ação Democrática Feminina Gaúcha (ADFG)*'.

13 Rede Brasil is an environmentalist watchdog Federation that follows the work carried out under the banner of IIRSA, founded and coordinated for a number of years by a NAT Brasil activist. IIRSA is the Initiative for the Integration of the Regional Infrastructure of South America.

14 Ironically, this process that Castells (1997) predicted as being inherent in the process of globalsation has been adopted as policy by the current UK government. David Cameron's 'Big Society' plan sees responsibility for certain public services being transferred to local volunteer groups.

15 Also see Doherty and Doyle (2009) on how European groups tend to consider their communities to be the cause of environmental problems, whereas South American, South East Asian and African groups consider the communities to hold the solution to environmental problems.

16 Many of these issues are tackled in a similar vein as in the work of academics such as Castells (1996) and Doherty (2006). Riles has argued, as regards network analysis, that the NGOs she worked with—and indeed, she claims, the NGO world as a whole—does this analysis for itself and that the work of scholars along these lines is redundant; it is the network seen twice. However, I believe that if one looks at the history of politics and academia, these have often mirrored each other. Anderson (2006) and Hobsbawn (1983) show how universities were the measure of a certain political consciousness—national consciousness—and that universities were the most vocal champions of these political orientations.

Therefore, it is the duty of the scholar to recognise what is not being asked, what is being assumed, rather than believe that the work of the scholar is in vain just because past work has spread out of the academy. It always has and it is in some way the hope of the academy that it should.

17 See Chapter 5 on how the activists' personal experiences, even from before joining a FoE group, are considered to constitute their FoE group and simultaneously FoE International.

5   Striving for an Exemplary Life

Becoming an Environmentalist

I invite you to imagine the following situations:

> *August 2007 Amsterdam, the office of the International Secretariat of FoEI.* Imee, the Members Coordinator and Stine, the Financial Officer of the Secretariat are discussing an annual report for Oxfam Novib, one of FoEI's main sources of funding. They go to Felicity's desk. Felicity is responsible for website and the member groups' details on it. Imee and Stine ask her how they had calculated that FoEI was made up of 2 million people. Felicity thought about it for a moment and then explained, "There are seventy-six national FoE member groups, from their reports we know that all the activists working directly with them add up to 1 million. Then from the same reports we calculated that those whom they work with indirectly—such as community groups and non-active supporters—would double that number to two million." At this point, they decided to reduce the number to one and a half million people, just to be sure they were not exaggerating. They talk about FoEI as being made up of the people, the sum of those who in some way work with or support a FoE group.

> *A couple of years earlier, in May 2005 in Malta.* Anthony, coordinator of FoE Malta, is preparing an application for a capacity-building project run by FoE Europe. There is a section dedicated to 'current capacity'. We are discussing this by back and forth emails as he writes it. Among other things, he includes as 'capacity': expertise on social justice issues, cultural and heritage projects, GMOs and a campaign on a proposed EU legislation on the registration, evaluation and authorisation of chemicals (known as REACH). Jurgen joined FoE Malta in 2003 and began to work with Anthony on the REACH campaign because of his professional employment as a food health and safety officer for a large chain of hotels in Malta. Once he joined the campaign team, it was FoE Malta as an organisation that acquired his expertise on chemicals. Jurgen's life experiences before and beyond FoE Malta were absorbed into what constitute FoE Malta. The same absorbing happens with Martina's MA in cultural management, and with Paula's work in social policy and my work in anthropology.

The office of FoE Brazil, October 2006, in Porto Alegre. Veronica and Amy have been working with the organisation for a number of years. They have come to know at least five volunteers who come to spend between six months and a year with FoE Brazil, just like me and Jennifer they explained (Jennifer was the other volunteer working with FoE Brazil while I was there). We are sitting talking in the front room of the office, surrounded by books and magazines, as normal. While Veronica works on her emails and Amy scours the daily newspapers, they ask Jennifer and me questions about where we are from and our previous experiences and interests. It is all informal and Veronica explains, "We chat like this with all of our volunteers from overseas. We have got to know so much about Germany, Belgium and England just by talking, even if we've never actually been. But most importantly we find out what you already know or like so your time here can be most fruitful for all of us." Because of our conversations, Jennifer and I were asked to give a talk during their weekly event lecture series about environmental issues and activism in our respective countries—Australia and Malta. Our experiences before going to Brazil become part of FoE Brazil's collective knowledge and experience.

What these three vignettes point to is the way that people, their actions and their knowledge, both past and ongoing, are considered to constitute the various FoE groupings. Part of the expertise for FoE Malta's REACH campaign came from Jurgen's work expertise. Yet the actual knowledge and practises he uses in his work never directly featured in the REACH campaign, although one day it might be. Similarly, in the Secretariat, activists in national member groups such as Jurgen, Amy and Veronica, who are not involved with FoE International, are counted as being part of who constitutes FoEI. The knowledge and actions of these activists are also claimed as being FoEI even if the activists themselves do not explicitly intend this. When they refer to persons, this includes their ongoing actions as well as, sometimes, their pasts, on the understanding that such experiences are present in the knowledge and kinds of persons they have become. Experiences and actions are considered to constitute persons; persons and their actions constitute FoE groupings.

Life before FoE is important to whether a person who wants to join a FoE group will be allowed to join or not. FoE Malta and the IS interview applicant volunteers, sometimes even twice, before they are accepted, even though candidates are scarce. Sometimes interviews are very informal. At my first international meeting, I stood up and introduced myself as a representative of FoE Malta as well as being there for my doctoral fieldwork on FoE. Tony Juniper (his real name), then director of FoE EWNI (England, Wales and Northern Ireland), spent two days of this week-long meeting coming to talk to me in coffee breaks and mealtimes, informally discussing a variety of things and eliciting my position on these things, as well as my past work. On the third day, it seems, I had passed some sort of acceptance point and my presence was not questioned again.

These different sorts of conversations, informal chats and interviews rarely include discussions about family. It is not imagined, as in many kinship-based groups, that personhood is defined by kin relations (Bamford and Leach 2009). The understanding of a person as an individual who has agency *as* an individual has been termed "agentic individuality" and is closely linked with philosophies of "civil society" in European thought (Seligman quoted in Hann 1996: 5). Okely (1996) shows that this "agentic individuality" is implicit in the form of biographical narratives. For FoE activists, it is the biographies and life histories, the actions, choices and positions held by the 'individual' that count.[1] In this chapter, I trace the life of one particular activist as he talks about himself, especially how he understands his own path to becoming a member of a FoE group.

This chapter focuses on the lives of the activists as they talk about themselves and the lives of others. In these conversations, what becomes evident are the resonances the activists' develop in their sense of self by working together: as a path to be actively forged through ongoing work and self-discipline. There is a discipline involved in continually bringing this self about. The activists' self-understandings, as I present them here, give a sense of how people with very diverse backgrounds become environmental activists.

Throughout the chapter, I hope to evoke a sense of what living an 'environmentalist' life is like, specifically for FoE activists. I bring my ethnographic material into conversation with Giddens' (2009, 2008, 2007) work on reflexive modernization and late modernity. Interestingly, the terms he uses resonate very closely with the notions and experiences of FoE activists and other environmentalists (Shepherd 2002; Moisander and Pesonen 2002; Horton 2003; Macnaghten 2003; Connolly and Prothero 2008). However, I suggest that Giddens's (2007) model of increasing globalisation and individualisation of selfhood are only partially useful in understanding how FoEI and other FoE groupings hang together.

The similarities that join the activists in their sense of personhood are elusive. During life history interviews or informal conversations, themes of personhood were not addresses in the same terms, nor did respondents describe exactly the same understanding of personhood. To fit all the different understandings into a unitary category would be a modernist method of analysis—generalisation with the aim of creating universals (Tsing 2005), a particular project of scale making that FoEI is implicitly combating.[2] Therefore, to the contrary, allowing even minor differences to count makes it possible to recognise how, in contrast to what Yarrow (2008) argues for NGO activists in Ghana, the FoE activists are not reproducing or expressing an underlying template or form that is legible independently of the specificities of each activist's life.

Rather, the similarities between them are more like musical counterpoint, in which different notes when played together (or, in this case, thought about or written together) overlap in a way that creates a distinguishable field. In fact, the activists themselves, in the life history interviews that I will explore in this chapter, speak of how they felt a resonance, or how they felt at home with the ideas and attitudes they found when they joined their FoE. For instance:

FoE's idealism mirrors my lifestyle.
                    —Diana, Finance Assistant, FoEI Secretariat

I began as a volunteer with Bettine, when I was reading things at FoE
and [snapping her fingers] things began connecting with my past.
        —Emanuela, Campaign Coordination Assistant, FoEI Secretariat

When I read the FoE Malta vision which I had not participated in making, I felt they were speaking for me too.
                    —Angela, activist, FoE Malta

Very similar in principle to counterpoint, where a musician needs to listen
to how different notes sound when played together, the overlapping can be
thought of as a process of co-respondence,[3] alluding to a sense of 'responding together'. I understand co-respondence to refer to different persons
responding within a common ground of experience (Ingold 1993, 2000b,
2002; Latour 2003).[4] The distinct field of FoE activism results from the
ways the activists co-respond to the affordances (Ingold 1992, 2009) of
that common ground. Affordances are not fixed characteristics, but emergent relational qualities, in which things and persons are mutually constituted within a 'constantly unfolding field of relationships' (Ingold 2000b).
Thus, co-respondence begins to explain how it is possible to have *different*
forms of environmentalisms, perceptions of the environment, and ideas and
practises of self, in a Federation that hangs together across so many countries. The different positions and differing relationships that constitute the
activists, and which they participate in constituting, resonate with common
aspects of a milieu and emerge differently depending on each activist's particular life history.

Affordances and Co-Respondence

The work of activism involves coming into contact with institutions, ways
of thinking and ideas that have been associated with 'late modernity' (Giddens 2008, 2007; Beck 1992, 2007). In Chapter 1, I showed how central
the nation-state is to the work of FoE activists. In addition, the mass media
and new communication technologies have also played a part in bringing
knowledge of similar environmental issues to the notice of FoE activists in
different parts of the world. FoE activists also co-respond during their own
international meetings and then through the stories participants take home
with them. However, there are other instances that are more serendipitous.
Many activists have been inspired by the same people. Anthony, the FoE
Malta coordinator, was inspired to include justice in his approach to environmentalism after having heard Vandana Shiva speak in New York. However, Andrea, FoE Brazil's coordinator, was also inspired in a similar way by

having heard Vandana Shiva speak in Germany, although Andrea focused more on eco-feminism as well as broader justice issues. Even the communications coordinator and long-term activist at FoEI IS had been inspired by Vandana Shiva early on in her life.

Their differences, on the other hand, emerge from the histories *they are*. The activists' ground of experience overlaps only in certain places. None of the activists I worked with had read Rachel Carson's *Silent Spring*, although it is quoted as a central book in writings on environmentalism. Emilia, an activist with FoE Brazil, insisted that I read *The Tao of Physics*, by Fritjof Capra (1975), a Czech physicist and popular writer. She explained that her lecturer in Agronomy had encouraged all his students to read this book, and it was this book that influenced her particular approach to environmentalism. In NAT Brazil only Andreana, Emilia's co-worker as well as classmate at university, had read the book. Anthony in Malta had heard of it but had not read it, and I passed it to him and other activists in Malta after having learned about it during my time in Brazil through Emilia. Milton's (1996: 218) notion of translocal discourses needs to be complemented with knowledge of the concrete pathways such ideas follow. This is so even if the paths themselves are mobile, in the form of (for instance) books, which are available anywhere reached by Internet bookstore services. These translocal discourses are not free floating. Tsing (2005) shows how even where similar stories reach different places, they can be re-shaped by local activists' own understandings. In Indonesia, among the nature lovers she worked with, the Chipko Movement in India was fused together with Chico Mendez's struggles in Brazil (Tsing 2005).

Finally, another aspect of co-respondence is that the activists *choose* to carry on co-responding in these shared millieux. Their co-respondence creates a certain degree of resonance, through which FoEI and other FoE groupings can be understood to 'hang together'. Although 'resonance' commonly elicits ideas of 'harmony' or 'communality', negotiations and contestation are essential to the notion of co-respondence, as I have come to experience it by working with FoE activists.[5] When an activist draws away from the milieu in which they can co-respond to FoE matters, as happened to me in the process of writing this book, many aspects of life change, as the grounds of experience change and no longer necessarily have enough in common to engender belonging to FoEI. FoE activists do not share an identical ground of experience, but do have enough overlapping common ground to co-respond with each other as well as with other aspects of their environments.

Choice has been understood as a part of agentic individualism tied to modernity and late modernity (Meyer and Jeppersen 2000; Connolly and Prothero 2008). But personal action is not necessarily personal *choice*. For instance, inmates in a Papua New Guinea prison smile or wink because a relative back home is thinking about them (Reed 1999). The action is personal, but they have no choice. The different ways in which choice is enacted by FoE activists suggest that assigning 'choice' exclusively to individualism is too narrow an understanding. I have argued that even if activists' senses of self co-respond, they are not identical. Therefore, the activists' ways of life cannot be explained along the lines that environmentalists are disciplined about their life choices because tradition, history and memory have been evacuated by "critical doubt" of high modernity, as Giddens (2007) suggests.

In what follows I show instances of co-respondence in the environmentalist lives of FoE activists. Their environmentalist activism is particular and is a result of the rigour with which they practise reflexive monitoring—to the extent that I characterise it as a discipline.

Agentic Individuals

The life history, as a method, has been criticised for concealing difference rather than making it apparent. The concern is that a life history interview assumes a particular sense of personhood (Okely 2001), the agentic individual, whose actions or biographical information, it is assumed, reveal something about the self (Abu Lughod 2000). In life history interviews, people who may not experience their lives in these terms are forced to adopt a subject position by the types of questions asked. While keeping these limitations in mind, Yarrow (2008) argues that interviewees or narrators are interlocutors who have their own motivations for responding. He argues that life history interview responses are performative; they are intended to have certain outcomes (*ibid*).

Is this the case with FoE activists? When I proposed that I would like to carry out life history interviews with people in NAT, the IS and FoE Malta, the response, apart from being positive (and busy people were willing to participate in interviews that sometimes last for up to nine hours), was in itself revealing. With NAT, I was gently reminded several times, by different people, that in Portuguese these sorts of interviews are called *trajetorias de vida*, literally 'life trajectories'. The NAT activists were familiar with life trajectory interviews.[6] I was not introducing something new by proposing to carry out life history interviews. Rather, even while I was proposing to carry out the interviews, they were already being partly reshaped by NAT activists as 'life trajectories', a concept of linearity I had deliberately left out of my explanation of the interviews. The understanding of life as a 'trajectory' or path emerges as one of the most prominent themes in all their life histories.

Activists are familiar with the form of the curriculum vitae (CV), which in its form crystallises and makes visible the notion of an individual's path of activities. Many of the activists I interviewed explained how, whenever they wanted new employment or to find volunteering work, they would send out their CVs. These CVs are often tailored to the particular job or work that is being sought, and different sorts of biographical information are included depending on the organisation to which they send the CV. In other words, the activists were not only familiar with, but also skilled in, the form of narrative implied by the notion of agentic individuality and so obviously in evidence in CVs. However, during the interviews themselves, the form of the CV was never kept.

This chapter and Chapter 6 are based on thirty-three life history interviews. Twelve with NAT activists, nine with FoE Malta and twelve with the IS. These interviews included working on or talking about people's CVs, daily work maps and participant observation, as well as very open-ended conversations about life experiences. The life history interviews ranged from a minimum of two hours, to a maximum of nine hours (in 3–4 hour stretches). The average duration was four hours, where the interviewee made a meal for us halfway through and the tone of the conversation changed over the meal. The participant observation I am referring to occurred especially at FoE international meetings or in situations where people who were not well acquainted were meeting. In addition, there were situations where I was introduced and got to know people individually throughout my fieldwork, especially the two volunteers I lived with, Jennifer in Brazil and Susana in Amsterdam, with whom I spent many hours talking about our lives.

Although my questions often brought the conversations that we referred to as 'life history interviews' back to experiences, actions and chronology, among other things, my interlocutors mixed memories with current feelings or situations. Often, distant memories were experienced as so present that on many occasions people were surprised or moved by how present their feelings or perceptions of those memories were. Dates and chronological order were never kept. Conversation flowed swiftly from a distant memory to how it is relevant to their lives today, to a happening or experience or a person this in turn reminded them of at another point in time, to hopes for the future.

Yet in conversing about their lives, the activists did enact and enliven agentic individuality. The activists discussed their lives in terms of activities or experiences they had had as individuals, not only during the life history interviews but also in conversations with each other. At international meetings activists who had not met before would get to know each other by exchanging information about their lives, and exchanging biographical information. As in a biographical or historical account where there are 'events', some actions and choices are considered more significant than others in how they constitute the person.

Biographical information is important on many levels: for reputation, for working out what new volunteers or employees will work on, and for getting to know each other (for the premise upon which 'work' relations are made easier, see Chapter 7). Since the sense of self conveyed in biographies and autobiographies was narrated by FoE activists in many such situations, I follow Yarrow (2008) in including these other situations in my understanding of life history and in the chapter.

Therefore, in this chapter I will present the narratives of self of six activists as they reflect on how they joined a FoE group.

A Life Trajectory[7]

Anthony is currently the coordinator for FoE Malta. He has been one of the core activists of FoE Malta since 2001 and in 2005, after Julian Manduca's unexpected death, he became the driving activist of the group. He began participating in FoE international meetings—both FoE Europe and FoE International in 2002. In 2004, he was elected to the FoE Europe Executive Committee, first as an alternate, then as a full officer in 2007. In this narrative, he explores how and why he came to join FoE Malta.[8]

Anthony grew up in Malta's capitol city of Valletta. Until he was seven, he explained that he 'only knew the city'. When he was seven, his family began going on summer holidays.

> We used to go on family holidays to Bahar ic-Caghaq. Friends of my parents have a large house there. *Dak iz-zmien kienet l-ahhar dar qabel l-ghelieqi.* [In those days, it was the last house before the fields.] We used to hang around, go for walks with our parents, their friends and their children. Their son was John Gatt, *you know him?* He's active in AD[9] and SSCN [now Nature Trust]—he was ten years older. He knew plants and birds and he used to show them to me.

Anthony emphasises that these childhood walks with John were the beginning of his love for knowing about the world. Kay Milton (2002) argues that environmentalists share experiences of loving nature in childhood. She recalls her own summer outings with her father, who would point out the different insects and plants they encountered on their walks. In the previous quote, Anthony relates a similar experience to Milton's. Both Anthony and Kay had these experiences as part of their summer holidays. As Anthony tells us further on, he still seeks places considered less inhabited by humans to find rest and pleasure. "Places with no paths", he tentatively calls these, "a little more wild". But his is not a simple appeal to a 'nature' that is opposed to culture, to humans.

While Anthony and I were on a hike, following what seemed like a forgotten path, certainly somewhere neither of us had been before, we came across a landscape criss-crossed with the remains of terraced fields. The fields lay in between and across the fallen boulders of the typical scree coastal landscape of the northwest of Malta. The fields were not only marked with dry-stone walls (*hajt tas-sejjieh*), but as they are intended to do, the walls catch the soil in the rain as it flows through them, building up soil and shaping the ground underfoot. Anthony was awed by the presence of fields so close to the sea, but we agreed that the place was not any less wild because of this trace of intense human activity. The wildness issued rather from the contrast between the business and stressfulness of daily life and the *leisure* associated with walks, hikes and camping. To return to Anthony's life history narrative, I want to emphasise that although I had asked about his childhood memories, his attention did not remain fixed in past recollection. In remembering the influence John had on his interest in flora and fauna, he related this to current situations—that John is still active with environmental politics and NGOs. He also asked me if I know him, not only relating this memory to the present, but also drawing me into his life narrative.

Anthony also positions his father as an influence. In the interview, he spoke of his father with admiration. Implicit in the way he relates the 'influences' in his life, Anthony connects aspects of the person he is today with the actions, choices and attitudes of his parents.

> My father was a policeman initially. He began working in the police HQ in the times of Mintoff. *F'dak iz-zmien in-nies kienu jidhlu fid-depo, u kien hemm min minnhom ma kienx johrog fuq saqajh, kien jaqla xeba tajba.* [In those days, people would come into the depot, and some of them didn't leave standing on two legs, they'd get a good beating.] A group of policemen decided to walk out, quit their job around '81. He lost everything then. He lost his pension and the possibility of early retirement—after 25 years of service. They all ended up working at Malta Post. He had wanted to stay in the police force but they left for political reasons. . . . When my father worked at the Malta Post he began a side job volunteering with a group of disabled people who make Christmas cards. He used to help collect funds and help out. . . . He still does.

Anthony relates "political reasons"[10] to notions of service and sacrifice; to follow his political conscience his father "lost everything". Anthony narrates his father's 'political reasons' in moral terms: the refusal of his father to participate in an increasingly violent institution.

He describes his father through the *actions* he carried out: leaving the police, beginning to volunteer, still doing it. As we will see more clearly in Anthony's words further on, a person takes charge of their life through the actions and activities they engage in. Over and above being able to change and *make* one's life through one's actions, Anthony judges taking charge in moral terms; he regarded his father a being 'right' to sacrifice his pension and a job he enjoyed in order to follow his conscience. Anthony links, in very close proximity, a sense of hope and getting on with things with his father's volunteer work. His father lost everything, but in his new job he sought other ways to give service, to "help out". To refer to oneself as an activist, or "to be active", in terms of this notion of agentic individuality, is then to have become fully a person. To carry on doing in the face of sacrifice is to further refine one's fullness as a person.

The sacrifice, service and the morality of living a life according to these notions influence Anthony more as an attitude than necessarily as a specific set of moral guidelines. Anthony's 'philosophy of life', while still being about service and morality and acceptance of sacrifice, is not limited to human beings, but is rather about openness and curiosity:

I used to like all humanities and sciences except business studies. I had the impression that they were subjects related, or based on, or revolving around human beings or rather services revolving around human beings. Whereas sciences are out of curiosity; this is the basis from which I function. Curiosity is what drives me. . . . Brother David, who now is the headmaster [of his secondary school], he's interested in everything. He set up the sailing club, astronomy club, chess club . . . he knows something about everything—you can say he is one of my role models. I sort of dislike people who are expert in only one area and not knowing anything else (like my boss) [he says with an impish smile, we laugh].

Encouraged by his parents, who both volunteered in NGOs, Anthony was interested in doing many things. At nine years old, he became a member of the Malta Ornithological Society (now Bird Life Malta) and later the Society for the Study and Conservation of Nature (now Nature Trust Malta). He read their publications and went to their events, like "raptor watches" and hikes. He is still a member of both. In the same attitude of curiosity, he joined the chess club and the astronomy club in his secondary school when he was eleven. A few years later, he joined the national Astronomy Society, of which he is still a core member. He has edited the society's newsletter for more than ten years. The Astronomy Society was an avenue that fuelled Anthony's curiosity and created more opportunities to explore different possibilities.

I knew I wanted to do something related to science, my friends said they wanted to be a vet—but I never had that [I wasn't sure what I wanted to do]. After O'levels I had a long summer using Astronomy society as an excuse—we [Anthony and two of his friends] would go to the bus terminus and randomly choose a bus number and stop where you stop. This was mostly at night, because we were astronomers. . . . We would walk until we reached somewhere—we didn't try to navigate . . . we would walk until you didn't have an idea where you were. We knew that Malta was not too big to be lost for too long. Astronomy taught me a lot about perspectives . . . huge universe, small world, small country. We knew if we were going North or West by the stars—but didn't know if a road was going to appear. . . . We had two rules: 1. Never turn back or if you've been there today don't go again. 2. If we were scaling down a valley side and it was getting complicated we would throw our tripods down as an incentive to go get it . . . the bumps and dents on the tripod we called them "medals".

Anthony's choices are guided by the desire to create more possibilities, but planning for the future is still central to how he makes choices: he plans for flexibility.

During his two years of junior college, aged sixteen to eighteen, he became an elected member on the Committee of the Astronomy Society. It is with the Astronomy Society that Anthony begins to talk of participating in the NGOs as being 'active'. Being 'active' has specific implications; being active means being part of the organisational core. He explained how, though he is still a member of Bird Life Malta and Nature Trust, he still receives and reads their publications, he even still goes on hikes and night hikes, but since he is not part of organising these events or activities, he is no longer 'active'.

In talking about the studies he chose to do after junior college, his reflexive planning for flexibility, to allow future eclectic choices, is even more evident:

> I took a year off after JC—I was thinking about what to do next—I was hiking a lot and doing astronomy. . . . It was still unclear [what I wanted to do] but I started radiography at medical school—it was a university course. . . . When did I start? Oct 1996 I think, or Feb 1997. I enjoyed the course, the technological bits were interesting and human anatomy . . . But I started thinking that when I finish the course I could only work for [the national] hospital—the private hospitals open were employing only experienced radiographers. I didn't want to be trapped.

He left the radiography course to undertake an Advanced Diploma in Computer Science (national computer courses instituted by the UK), run in Malta by a government agency but offered as distance learning by the University of Nottingham. He completed this course in 1999, but before he told me when he finished his course, he talked about building his family tree, his dislike for remembering dates, and he explained further how he plans for flexibility. Anthony's reflexive self-planning is guided by his ideal of a 'curious' person, eclectically open to different sources of experience:

> I like opening the door for opportunity to happen, so you don't force the situation, you don't need to focus your energies and time . . . but mentally being open to things, to change by doing more of other things—to leave channels open. Maybe I don't say no sometimes, I say "let's try it", "let's go for it and see what happens".

I asked him: how can this be reconciled with planning for a sustainable future? He was challenged by this question. He thought about it and said,

> It might be less disorganised than it seems—my planning doesn't feel random. For example, in the Statute we are writing for FoE Malta, I am not thinking about myself as a coordinator but about an ideal. I do think about specifics but also generalities. I can say that this general thinking is linked to astronomy.

Flexibility, eclecticism and curiosity are the generalities which he plans for himself and in the hope that others (both human and non-human) will have those choices too.

Chronologically, in his life history narrative, he has not yet become a member, nor even met FoE Malta, but he ignores chronology and reflects on the difference between FoE Malta and The Astronomical Society of Malta, on 'politics' and his 'environmentalism':

> The Society is mostly different to FoE. The Astronomical Society shuns politics, they try not to get involved in polemics. . . . Yes there are some things that are changing, the issues about light pollution mainly, but they are non-confrontational, they prefer to work through advisory tactics. . . . On the Board they don't have the sorts of discussions we have in FoE . . . I don't bring up political issues myself in the Society. . . . Because I want to get a break from politics and they have a different way of doing things. The Astronomy Society is more of an educational approach. FoE simply *is* political.

In this passage, we begin to get a feel for what is particular about his 'environmentalism' with FoE Malta. Politics for Anthony is about engaging in open debate—"polemics"—in the same way that he described his father's leaving the police force as political because his actions were in open disagreement with the practises of the state.

Here his analysis of the "non-political" work of the Astronomy Society as being 'a different way of doing things' points to a central value in the 'vision' of FoE Malta and FoE International. His acknowledgement that there are 'different ways of doing things' reveals not an attitude of tolerance towards others, but a valorisation of different strategies. Anthony is 'active' in *both* FoE Malta and the Astronomical Society without considering that his non-confrontational approach in the latter contradicts his political environmentalism in the former.

His reflection on the difference between the two organisations led him to tell of how he met the woman who introduced him to FoE Malta:

> End of October in 1999 I met Hannah at an Astronomical Society night hike and we talked all night about vegetarianism and environmentalism . . . Hannah was here in Malta till March 2000. At the time I wasn't a vegetarian yet—but I already didn't used to eat a lot of meat. I already did consider myself an environmentalist from the hikes we did—I was very aware of land issues. There were so few places you could go that were not full of building sites. . . . And there were the hunters. They would throw stones and things to stop us walking in the area [where they hunted], but we didn't give a shit—if anything we went back at night and removed their nets [the net hanging from the ceiling of his room was from one of those night hikes], we would demolish huts,

remove the poles for *nsib* [bird trapping with nets] . . . Even with the astronomy group we did this demolishing. They are also aware of being a small different planet and it's about appreciating what we have—it's more a conservationist approach . . . Hannah was organising an event at University about compost bins for FoE, this is where I met Julian for the first time and Gregory. Gregory came with a van with the boards and I went to help put them up.

Of Maltese parents, Hannah was trained in environmental law, and worked on recycling and composting projects for the local council where she lived in Long Island, New York. Exploring the possibility of moving permanently to live with Hannah in the United States, Anthony left his job in Malta and moved to New York on October 28, 2000. 'I remember *that* date,' he smiled. From his experiences in New York, Anthony singles out a specific event as a direct influence for him to join FoE:

> There was a conference at the New York City Uni about technology and globalisation—lots of workshops where Vandana Shiva was the highlight, she is like the Mahatma Ghandi of today. It was amazing actually, the way she spoke, it was so powerful. . . . Environment and social issues was the evident thing, like the main point of the conference. . . . She was talking about if you don't want to depend, you know due to globalisation you have to bake your own bread and weave your own cloth. . . . There was another speaker on the US space programme and militarism, I used it to write an editorial for the Astronomy Society newsletter. My editorials are very political. . . . This conference inspired me to come to work with FoE when I knew that I was coming home to Malta.

Before moving to the United States, he had sent his CVs by email to various NGOs and companies in the United States. He was accepted 'to work in a company that was an NGO, it was computer work. They needed a lot of Italian speaking workers like myself'. Nevertheless, he was having difficulty getting a work permit, so he decided to return to Malta. Planning for flexibility requires the creation of opportunities. As Anthony says, the flexibility and eclecticism of this planning of one's life is not about doing things at random, it is about making allowances for change within a particular ideal of life. Like most of the other FoE activists, Anthony searches out opportunities for work or activism before significant changes such as moving to a different country. He returned to Malta with a job lined up with an online betting company and a plan to be 'active' with FoE Malta:

> I came back from the US and took a Green MEP to Tal-Virtu with Julian. The MEP was Friends of the Earth Luxembourg, that's how Julian knew him. Sometime that day Julian interviewed me and in 2001 I worked on the Agriculture project [a report funded by MEDNET on

biodiversity in agriculture]. I did the research, the writing, and Julian and Katherine were proofreading. We did a lot of work by email.

In Anthony's narrative, once he reaches memories in which he considers himself a core activist, he switches much less between different tenses. It is possible that this happens because these would be the stories shared with other FoE activists and are therefore more structured by virtue of being more practised, better rehearsed (see Yarrow 2008).

> At the time the Verdala golf course campaign was still going. We were going to meetings and a public hearing, it was my first public hearing.[11] It felt like going to war. Anglu Xuereb brought all his workmen—they were causing havoc. There were foreign people—one of the people carrying out the EIA [Environment Impact Assessment]—so they began the meeting in English—everyone shouted and swore about it—we didn't want the foreign EIA people to speak, especially not the farmers. Most of the farmers didn't even understand English that well. That's when I spoke back to Anglu Xuereb. It was quite controversial. I brought up the pesticides in his golf course: "do you want to fill up your hospital with people suffering with cancer to fuel your own business?" This was quite mild in comparison to the comments we were receiving from them.

As Anthony describes it, his confirmation in environmental politics was the moment when he "spoke back". Similarly to the environmentalists with whom Eeva Berglund (1998) worked in Germany, Anthony could speak back based on his knowledge, on his research into agriculture and pesticides. His knowledge included the effects of pesticides on the health of humans and non-humans. Knowledge, books, science and both formal and informal education feature strongly in all the life history narratives I collected. The FoE insistence on including humans and justice issues seamlessly with other environmental concerns translates into an attitude that seeks to value difference over homogeneity in group-constituting practises; it shapes activism into the experiment in inclusive decision-making processes that is characteristic of the FoE International meetings.[12]

As with Anthony's life history, the understanding of personhood as agentic individuality, as life being an individual path, is present in some form or other in all the life history narratives of core FoE members. Being 'active', not necessarily an 'activist', nor an 'environmentalist', but active in *choosing* one's lifestyle, is a central theme. More specifically, being active about one's life plays itself out as planning for flexibility. As Anthony noted, this is not about being 'random' but about opening up the possibility of choice. From the life history interviews, it appears that the notion of life as a path is reflexive and managerial. As such, it can be understood as a technology of the self (Foucault 1988), steering one's life as one moves along. In Chapter 6, I elaborate on these themes, which resonate throughout the life history narratives of FoE core activists.

Paths and Trajectories

FoE activists talked about life as a path. Each person has their own individual path. This life path has to be made by the individual; it is not set out for them in advance and depends on their actions and choices to be carried forward and followed. Various aspects of life history narratives, including the life history interviews and the Brazilian name for them ('life trajectories'), exchanging biographical information at conferences, interviews with new volunteers and most starkly the ubiquitous use of CVs, help to solidify the coherence of this path as belonging to an 'individual' (see Linde 1993; Ochs and Capps 1996). This path (solidified through various means, including the narratives themselves) is actively forged by the individual and is part of the activist's self-commentary.

Noemi works on forest projects with NAT Brasil (Nucleo Amigos da Terra/Brasil—Friends of the Earth Brazil). During her time at university studying Forestry (*engenheria florestal*), she discovered that she could use her knowledge to work in the environmental sphere; to do something, she explained, about the things she worried about. Having graduated, she returned to her home in Porto Alegre to do a master's in Ecology at the Federal University of Rio Grande do Sul (UFRGS). In this two-year period, she returned to live with her parents. She described this move as a shock; she felt she had to readapt. '*Sou uma pessoa que se pode adaptar, demora um poco*' (I am a person who can adapt, even if it takes me a while). And she can adapt because: '*tem uma meta, pra alcancar a meta*' (there is a goal, to reach a goal). 'Consiguo durante 2 anos, durante 10 não' (I'll manage for two years, for not ten). Getting her master's was '*uma meta alcançada*' (a goal reached). '[*Tenho*] *uma vida cheia de metas*' (I have a life full of goals) she joked. After her master's the next goal, in order to do something about 'the things she worried about', was to join an NGO. '*Foi procurar NAT, procurar informações sobre Cambara do Sul, campos ensima da serra* (places where she used to go on holiday). *Dai nunca mais sai* [do NAT]' (I looked for FoE Brazil, I found information about Cambara so Sul, the mountain plains. From then on, I never left FoE Brazil).

Emanuela grew up in Argentina, but travelled every year to visit her Peruvian mother's family just outside Lima. In the life history interview, she described what she saw as a stark contrast between Argentina and the Peru she encountered. In Peru, ethnic and class difference were apparent and disturbing to her. Because of the many injustices, she says this experience alerted her to, she decided she wanted to study sociology at the university. The departments of sociology and philosophy were closed during the military period in Argentina and opened only in the year she applied to go to university. 'I wanted to be effective,' she explained, 'but it was a chaotic time and sociologists were known to bring about change.'

> I also became more Marxist at that time. From a conference I went to and a number of books that I read about the situation in Nicaragua

I began to read about Marxism and Che Guevara. I read the biography of Marx for example. I wanted to be an intellectual. I read "Materialismo dialectico" written by Emanuela Hagnicer—a woman reviewing Marx. I read Althusser too. . . . If I studied law I would understand the superstructure—the needs. Then I would understand culture, values: sociology would come second . . . I wanted to study in Texas, which I knew about because many international people came to live in my mother's guesthouse. LBJ uni in Texas had a Latin American studies department—it has a very strong Latin core. That was the strategy.

Emanuela is reflexive not only about having a life path or project, but also about the source of that particular project in her readings and encounters with particular people.

Reflexivity is valued in environmentalism. A central principle of environmental management is "monitoring". Having a life path, and which path, is part of the same principle as monitoring. During my apprenticeship in FoE Malta, I increasingly found myself monitoring my choices and actions according to a vague notion of sustainability. Having been shown a film about the violent workings of the Coca-Cola company in Colombia, I began choosing not to drink Coca-Cola (the choice was an internal battle every time I bought a drink). Later still, after more reflection and discussion, the practise turned into not buying or drinking fizzy drinks made by multinational corporations. Mariangela in Brazil uses the drinking of Coca-Cola as an indicator: if they drink Coke, they cannot be "good" FoE activists.

Eclecticism and Flexibility

FoE activists understand personhood in terms of an agentic individual who not only makes his or her own path but also, through the critical monitoring of lifeways, has a particular technology of the self. In Foucault's words, a technology of the self allows "individuals to effect by their own means, or with the help of others, a certain number of operations on their own bodies and souls, thoughts, conduct, and way of being, so as to transform themselves in order to attain a certain state of happiness, purity, wisdom, perfection, or immortality" (Foucault 1988: 225). The forms these technologies of self take can be differentiated according to three characteristics. First, the active constitution of the self is an administrative, progressive project, not a juridical, punitive end-point (ibid: 34). Second, the self is constituted through *practises*. The term *askēsis*, central to Greek technologies of the self, denotes the training of oneself by oneself (Rabinow 1984: 364). These technologies include practises such as contemplation in early Christian monasticism and imagination in the Stoic tradition (Foucault 1988: 36–37). Third, techniques of self will differ according to the beliefs held or, in Foucault's terms, they will depend on a people's *telos* (Rabinow 1984: 358). FoE activists' reflexive monitoring of their life path corresponds well with the first two defining points. However, as we will see further on in this section,

FoE activists' *telos*, ideal or goal, contrasts with the disciplines examined by Foucault and others. The FoE core's *telos* is plural, eclectic and flexible.

In Anthony's life history, we saw how he planned for flexibility not only in the choices he made about what to study, what work to do and what choices to make, but also in the attitude with which he engaged in his activities, as epitomised by his night hikes. The hikes were shaped by two explicit, self-imposed rules:

1   Never go the same way twice.
2   Create incentives for facing up to challenges (throwing down their tripods).

And they had two implicit rules:

1   Catch any bus and don't plan where to get off.
2   Never take any maps.

These various types of self-imposed rules establish a discipline characteristic of a technology of the self-planning and monitoring one's life according to an ideal. A particular type of planning is needed for the flexibility and spontaneity FoE activists seek.

The active planning of paths and trajectories, the fact of having a reflexive technology of the self, is integral to FoE activists' personhood. Yet these paths are not straight lines, nor do the activists strive for, expect or desire a uniform, monolithic, unchanging or rigid path. It is understood that the coherence of the path comes primarily from the effort of continual movement, making and travelling a path. So what are the ideals that guide the FoE activists' technologies of the self, their disciplined lives?

In Amsterdam, I would cycle to the FoE office. I had my route; I went the same way every day. I left my neighbourhood and crossed the Amstel. I joined the dwindling crowd of others on their way to work, crossing tram lines, negotiating overtaking bicycles and cars, even imagining familiar strangers, people I seemed to recognise from travelling the same route each day. One morning, however, I did not head straight for the office. First, I went to Guntur's flat (Campaign Coordinator, FoEI Secretariat) so that we could ride to the office together, pick up some papers and go on to a morning session of Bikram yoga. His flat was not far from mine and I assumed we would eventually join my normal route. We never did. Instead, we wove in and out of parks, through passageways between buildings barely wide enough for the bikes, and emerged onto the main streets only a short cycle before the office. I had never passed that way before. I doubt that I would be able to trace it back if I tried. On the way, Guntur called above the wind that he always looks for new paths to get to work and to get home. Except for when he's late or has a meeting to get to, he tries to do this every day—even if it makes the route longer. He tries to avoid the main roads at all costs, but he also does it to discover the city. He will never use a map to get to know a new city he visits, but will walk and get lost until he figures his way around.

Another time, leaving the office, I cycled with Felicity, who takes care of FoEI's website. We were going to her flat to do her life history interview. Heading towards where I lived we did follow the main route, but stopped at the *naturwinkel*, the nature shop, on the way. Felicity walked through the aisles, stopping every now and again when something caught her eye. She did the same in the vegetable section and picked out some free-range organically fed chicken on the way to the till. She didn't have a list. Back at her home after a couple of hours talking about her childhood, she began preparing supper and started off by searching through her cupboards: 'I think I'll make pasta . . . I'll add capers . . . I have olives'. Interested in how environmentalists regard their own consumption, I asked how she planned her meals and her shopping. 'I like to cook what I feel like on that day . . . and the nature shop is on the way back from work, so it's very easy just to stop and pick out something I feel like'.

Anthony, Guntur and Felicity, similar to many of the FoE core, make space for spontaneity; they plan their choices and actions to allow for flexibility. Flexibility in and of itself, as sometimes with Anthony, may be the ideal. Flexibility is a shorthand for different types of eclecticism. It may refer to: discovering unknown places—Guntur's roving; allowing one's impulses to be followed—Felicity's cooking; or allowing new or unexpected experiences by being willing—Anthony's "not saying no" to give new experiences a chance. As with Anthony's hike rules, Guntur and Felicity also hold to self-imposed "rules": Guntur never uses a map and avoids main roads at all cost; Felicity will not plan the meals for the week ahead, and will look through what she already has to make her meals. These rules make the discipline; the ideal is held up and practises moulded to it. Whether the practise or the ideal comes first is not the issue, what makes it a discipline, a technology of the self, is the claim that they strive to achieve an ideal through their actions.

This discipline of flexibility is consistent with the sociological understanding of "identity" in the post-modern condition (Baumann 1995; Hodges 2008). In the rhetoric of post-modernism, identities are not ascribed; individuals have wrested themselves away from the authority of tradition and can make their own lives, drawing on whatever sources they like. "Postmodern identity" is described as eclectic (Baumann 1995). Without digressing on the question of whether I am describing an identity or something else (see Chapter 6), the FoE activists' lifeway, while recognisably a technology of the self, is starkly different from technologies of the self that are not eclectic. A comparison will highlight this difference.

Talal Asad and Peter Van der Veer both explore "disciplines of the self", based on canonical sources. Asad explores medieval Christian monasticism while Van der Veer analyses the different branches of the Ramanandi monastic order in Northern India. Benedictine monks in medieval Europe strove to attain virtue of the mind (developing the cognitive sensibility to recognise the "Truth", Asad 1987: 200), body (*in habitu, in gestu et in mensa* [*ibid*:170]) and soul which became refined by the attainment and application of Christian abilities (*ibid*:183). The mind educated the body and soul because it was seen as the vehicle through which recognition (remorse), penance and subsequently virtue could be obtained (*ibid*: 196). Similarly, Ramanandi monks discipline their body according to the beliefs of their particular monastic group. The *tyagis* suborder has a strict ascetic discipline aimed at accumulating heat (*tapas*). The *nagas*, on the other hand, dedicate themselves to a military discipline (Van Der Veer 1989: 462). The third subgroup are the *rasiksa* who are male but programme their behaviour by enacting the part of Sita's handmaiden and don feminine dress and the feminine attitudes of devotion and servitude towards the godhead, Ram (*ibid*: 466). All three orders are celibate to enhance their bodily power, which is seen to be diminished by sexual intercourse (*ibid*: 464). For both Ramandandi monks and Christian medieval monks the most important process at play is what Asad refers to as *disciplina*, where daily actions and reactions (or feelings) are minutely planned and monitored according to theology, cosmology and morality (Van der Veer 1989: 460). The monks' lives are a literal incorporation of belief (*ibid*: 459).

Medieval Benedictine monks willingly devoted themselves to re-education according to a number of established authorities (Asad 1987: 187). The Benedictine monks submitted themselves to the hierarchy not only of the monastery but also of the central Church. Ramanandi monks follow strictly regimented lifestyles that are based on a single version of the story of Ram. In both cases, there is a clear reference point. However, unlike the Ramanandi text and the Benedictine's submission to papal authority, there is neither central authority nor text for FoE activists. The lack of a canon or authority seems to be a recognised feature in environmentalism. Latour asks whether this canon could have arisen from the scientific discipline of ecology, but suggests that this is not the case since it does not provide a unified science (2003). Berglund (1998) also finds that certain environmentalists in Germany do not trust science because it is seen to be the cause, at least in part, of the ills they are combating.

The loose structure of the FoE International (FoEI) Federation is aimed at reducing the extent to which institutionalisation and standardisation can occur. The FoEI General Meeting in Beatenburg 2002, upon the suggestion of the FoEI Executive Committee, decided to change the Annual General Meetings (AGM) to Biannual General Meetings. The decision was an effort, among other things, to reduce centralisation in the Federation. Not only is there no solid central authority or text, but also and, to the contrary, the FoE national member groups are *expected* to have an understanding of "sustainability" that is particular to their national context and to make their claims for different understandings based on their particular historical and environmental experiences.

During her life history interview, Jennifer, a founding member of FoE Malta, described her discipline:

> Changing the way we do things is about taking responsibility on yourself, not like those people who can't see the consequences of what they do beyond the four walls of their houses. The difficulty is finding the cut-off point. For example we have just had tea, for which the water was boiled in a kettle, which used electricity, but I'm not going to stop having tea. It's hard to know where to set boundaries.

The point is that as environmentalists, the open-ended ideal does not prevent them from monitoring their actions and choices. As opposed to the explicit canons of the Church (Van der Veer 1989: 460) and the *Ramcaritmanas* text (*ibid*: 461), the sources from which FoE activists draw experiences and information to guide their process of self-disciplined "sustainable lifestyles" are eclectic and flexible. Veronica (FoE Brazil) reflects on her being an "environmentalist":

> People try to change but it's difficult. . . . Look at urban gardens, my garden has ants, what to do? The ants will kill that tree, do I kill the ants or let the ants kill the tree? . . . These earrings I remember that these diamonds might be the fruit of war . . . I guide my choices and actions by feelings, priority of feelings, not the rational mind. This is when the spiritual is important . . . I make decisions based on what my conscience can put up with [*o que a minha consciensia pode aguentar*].

Flexibility can be seen as necessary when the situations in which discipline is required present themselves in unexpected ways and from unexpected sources. Eclecticism can even be understood as co-producing the valorisation of diversity embodied in the decentralised structure of FoEI: not having a centralised, single canon (whether text or authority) is part of the discipline of flexibility.[13]

Wayfaring

Apart from the lack of a uniform source or single ideal to inform their technology of the self, planning for flexibility is also needed because the activists will have more than one set of ideals. Their other ideals, involving different technologies of the self, different disciplines, may overlap in the situations they face. Emanuela (FoEI Secretariat) had a strategy, her activist plan "to bring about change". However, she had other hopes and plans, with their own strategies.

Emanuela's strategy was to study law in Argentina and then apply to study in Texas, where she would follow the second stage of her plan—sociology. However, then she met Karsten. He was staying in the guesthouse her mother ran in their home, which was close to the university and a research institute with Dutch connections. Karsten was the sixth Dutch person to stay there. He was there to do the practical period for his geology degree. 'I told him don't flirt with me, because I'm looking for the final one'. She had graduated in law and wanted to study in Texas, to follow her "strategy",

> but Karsten was important. So instead we lived in Italy for half a year where he had gotten a job. Italy was like a Jean Paul Bel Mondo—a film—it was beautiful, winding streets. It was not what I was looking for. . . . We could have stayed five years, but I didn't like the mentality— very segregated mentality. . . . Image was very important, they were very conservative and consumerist—very like Argentinians. But Italians did not have the poverty. . . . So I could either go to Argentina where I already had networks from being an activist. Or I could go to the Netherlands to study development studies—that was my aim.

She got her papers to work in the Netherlands. There she lived with Karsten's family. His mother found her a job cutting lettuce in a factory, while Karsten worked in Italy. She went to the embassy to change her passport and asked what Argentinian lawyers did in the Netherlands. Two months later the Consul called her for an interview in The Hague. She remembers that the ambassador asked her about her political position. She said she was a Social Democrat but she was not supporting Cristina (Fernandez de Kirchner, current president of Argentina, a member of the Justicialist party, and a member of the Peron Youth in her adolescence). 'The Hague was full of Shell people, people connected with Cristina'. She said she learned from that what she did not want to follow. There too she found "lifestyle", meaning people interested in having luxurious lifestyles, similar to her experience in Italy. She worked at the Consulate for two years doing administrative work. She chose to go to the non-profit sector to "continue my strategy".

Like Anthony, who sometimes deliberately does not refuse proposals for the sake of following unexpected paths, Emanuela's strategy to bring about change, and her plan to have a family with Karsten, brought her into situations that did not match either ideal. Her having these two ideals created more possibilities of not being able to stick too rigidly to either one or the other. Veronica explains that she depends on her husband to do her work for FoE Brazil, because he understands the amount of travel her FoE Brazil work requires. Within an ideal of flexibility and eclecticism, different plans, ideals and strategies are not necessarily understood as impediments to a successful discipline of self. In fact, in their study of "green consumerism", Connolly and Prothero (2008) also find that environmentalists have to negotiate their discipline with themselves due to the different relationships that also constitute them; their being mothers and friends. They argue that although Giddens's (2007) and Beck's (1992) work provide useful frames of reference for understanding reflexive environmentalist practises in the context of pressures that promote increased individualization, the environmentalists and their practises are none-the-less socially constituted.

Kohn's (2008) analysis of the discipline of Aikido is in complete agreement with this position, as she shows how selves are simultaneously both construed as agentic individuals and constituted in relations with others. Mary Carruthers (1998) shows how the lives of medieval monks were socially and collectively reflexive and disciplined. These works show how reflexive discipline is not just a function of individualisation or of the conditions of late modernity. Although increased individualization is present in pressures and possibilities for empowerment (Connolly and Prothero 2008), the activists' discipline cannot merely be attributed their constitution as "reflexive individuals" in "reflexive modernity". Something more particular to FoE is happening.

In the way the FoE core members plan for flexibility, pursue multiple strategies and "adapt" (as did Noemi) to unexpected situations, their technologies of the self may be better understood as reflexive wayfaring. Wayfaring implies that 'the world is *not* ready-made for life to occupy' (Ingold 2000: 240). Each situation a person encounters is necessarily a new one. All life is an improvisation: in Ingold's words, 'we know *as* we go, not *before* we go' (*ibid*: 239). FoE technologies of the self, understood as reflexive wayfaring, are also about going along with life, but as we saw with their educational choices, FoE activists also plan for this flexibility. As part of the ideal of openness and flexibility, their plans and the means to achieve them are constantly being formed, changed or reaffirmed, sometimes articulated, all the while remaining a discipline.

Politics, Environmentalism and Class

In ethnographies of conservation, environmentalists are often portrayed as white, middle class, Western and generally men.[14] The central subjects of these works are usually the "non-elite", the "rural", or the "local" inhabitants, or a situation or place.[15] In contrast with these non-elites, environmentalists are often portrayed as imposing their view of nature, of foreclosing politics. Rabinow warns about the new kinds of disciplinary power environmentalists hold. Anderson and Berglund (2003) similarly warn that environmentalists may be the elites of the future. Little (1999) emphasises the covert tactics of environmentalism and Ferguson (1990) call this "anti-politics". Latour (2003) comments on how science and expertise in environmentalism are used to close off debate, while Cristina Adams (2003) shows how the politics of nature parks disempowers indigenous inhabitants. Environmentalists are pictured as a homogenised, seemingly unified, movement even though the few in-depth ethnographies that there are show how this is otherwise.[16] Other authoritative anthropological writings dedicated to environmentalism are not grounded in detailed ethnography (Argyrou 2005; Milton 2002; Ingold 2000a).

What seems to be at play in the identification of environmentalists as white, middle class, etc., is the assumption that identity is absolute, that if one is socialised as middle class in childhood, then one can only have middle class responses and understandings. This argument is encapsulated in two popular books *Mafia Verde* and *Ambientalismo: Novo Colonialismo*. These books published in Brazil were suggested to me by environmentalists I met in Porto Alegre as exemplars of anti-environmentalist discourses in Brazil. These popularly aimed publications are supported by scholarly work. For instance, Bourdieu's (1990: 52–66) statement that early experiences shape the perception of any future experiences is the scholarly version of the same assumption. My ethnographic work with FoE activists shows that this position veils how learning occurs throughout life (as the practise of anthropological fieldwork hopes and expects). Moreover, such notions, including Bourdieu's habitus, effectively deny the role of reflexive, conscious, articulate thinking (Farnell 1999; Starret 1995). Here the point is that FoE activists engage in self-conscious discipline which complicates any simplistic categorisation of persons as 'middle class'. Consider, for example, the way 'reflexivity' is mobilised by anthropologists in the effort to decolonise the practice of ethnography (Marcus 2001), no matter the background of the individual anthropologist.

In the previous section, I showed how FoE activists reflexively discipline their lives according to an eclectic and constantly changing ideal of sustainability. In their resonant technologies of the self, they hold in common the notion that humans are part of the "environment", not separate from it. Recall for instance Emilia (Chapter 4), who criticises volunteerism in environmental NGOs because this restricts the sort of persons who can contribute. The issues of justice, social inclusion and class imperialism are topics of personal discussion in this process of self-management as much as they are explicit points of debate in their collective FoE discussions. Since class is something they actively think about and question, it cannot be used to explain away the work and lives of environmentalists as though they were blindly self-serving or as varieties of neo-colonialism.[17] Many activists readily identify themselves as "middle class", and suppose that their relatively privileged position requires them to offer service to others, to work for the empowerment of others, both human and non-human.[18]

In relation to the first critique—that environmentalists, being white, middle class, Western and male can only reproduce the logic of their culture and class—I will show in this section how, as part of their reflections upon their socio-economic histories, FoE activists stress how they have learned from others throughout their lives.[19] They have learned from others whom they identify as either similar to or different from their socio-economic background. They have both radically changed their own assumptions or chosen to maintain them—both are reflexive choices. Finally, the life histories of some environmentalists, in this case FoE activists, are in themselves histories of socio-economic change. Not only is it possible for activists to understand

and work for others who do not necessarily share their cultural, historic and socio-economic position, but the socio-economic backgrounds of FoE activists are much more diverse than is portrayed in both popular and scholarly literature.

Emanuela, based in the Secretariat in Amsterdam, explained that she grew up in a very middle-class family: 'As a child I was forbidden to go out with boys, I had to study and to finish my work before going out—well we were a middle-class family.' That middle-class upbringing did not prevent her from recognising different ways of life. She talked about how she was 'aware of the political implications' of class and ethnicity and the injustices that surrounded the issues:

> In Argentina there is no tension. But in Peru I saw the tension. You see the tension in the model of beauty, in education. In my family, my mother is fair and my father is darker so father was considered Peruvian and my mother an Argentinian. I saw these injustices and poverty in Peru from a very young age.
>
> We used to visit in Peru and these indigenous people cleaned your house, they were treated like slaves. Whereas in Argentina the people who cleaned the house were treated like employees.

When she joined FoEI, at the age of thirty-four, she felt a resonance with her own political interests, but learned a new direction for her interests to take:

> I came to find out about environmentalism through FoEI. My country was destroyed by IFIs [International Finance Institutions] so I went to FoE still focusing on development and IFIs. When Bettine and I were preparing a document [on environmental rights] I realised that poverty was widespread and the only richness the people had was the clean environment. On the other side of their poverty they had access to water. So a clean environment is a human right.

Lili, who is also based at the Secretariat in Amsterdam, reflects on how she came to consider social issues essential to her environmentalism probably shows most clearly, the degree of reflexive questioning about learning from others with different socio-economic backgrounds, and promoting their interests as well as her own.

> Hong Kong was very materialistic, I knew a lot of spoiled kids, and there is a bit of money there but Dad didn't make a lot of money on a civil servant salary. Mum was middle class and had much richer cousins, so she brought me and my sister up to be aware of class distance—so Mum treated the housekeeper very well, things like not leaving clothes around for other people to pick up.

Mum took me to a peace and anti nuclear rally in London [Lili drew the peace sign on my notebook]. My parents weren't hippies—Mum became vegetarian for about ten years—but they were more mainstream. I get my political awareness partly from my Mum. But my godparents are very political/hippies, they listened to Bob Dylan—we knew them in Libya.

At university, her friend Amber (who was a friend from the secondary school that Lili attended) was what Lili called 'very political'. She went on to become a community worker, with a master's in community development, in places like Bangladesh. On the other hand, she had two other friends, Rachel and Matt, who were, to use Lili's words, 'extreme environmentalists'. Together with Rachel and Matt, Lili experimented with being vegan for about a year.

When Amber—she grew up in the UK—argued with Rachel & Matt I couldn't understand why they didn't make connections between the environment and social. . . . Matt is one of those very arrogant northern environmentalists. He grew up in the UK and just doesn't know southern countries, he thinks the environment should be put first. . . . There were shanty towns in Hong Kong—lots of Vietnamese refugees where I volunteered as a teenager . . . I was not particularly encouraged by my parents . . . my sister became an occupational therapist but already from before, she also had friends who came from missionary families.

I asked her whether she considered her sister to be political in the same way that she is:

I've thought about it before—why am I political and my sister not? A lot of my sister's friends are missionaries but I think it is quite patronising, not really political.

Veronica (FoE Brazil) grew up in what she called a lower middle-class family and never finished her secondary education. She often reflects on the debate about whether environmentalism should be mainly conservation or a rounded political project. As I described in Chapter 4, Veronica had disagreements with most of the other activists about working in partnership with private companies, more specifically with paper makers who manage tree plantations. Nonetheless, Veronica had enough influence in FoE Brazil to prevent two other activists, working on a project about eucalyptus plantations, from calling the plantations "green deserts".[20] However, Veronica was forbidden to hold official meetings alone with representatives of such plantation companies. She complained to me about this, saying that it limits the success of her work for the protection of the Atlantic rainforest. In the life history interview, she explained:

There are different people in the environmental movement. There are the political ones on one hand and the conservationists on the other—I see myself right in the middle of that spectrum—because I also try to conserve as much as possible.

There are a number of other core FoE activists who also insist on a conservationist approach. Historically, FoE Germany and FoE Switzerland are conservationist NGOs. Bruno, from FoE Switzerland and Angela, from FoE Malta, like Veronica, argue that even as FoE activists, someone should be pushing for "Nature" to be protected as distinct from human interests. And yet, though Veronica and these others shared early experiences of "Nature" and leisure, this does not prevent them from being reflexive about their conservationist attitudes, even if they then choose to retain their conservationist stance. While the FoE activists' background may be predominantly middle class, they have come into contact with people from many different backgrounds and experiences that are not part of their "middle class" upbringing. That has changed their understanding and influenced their choice of lifestyle. They are not trapped in their childhoods; they comment on how they are different from their siblings and their parents, although influenced by them too.

I hold that the important aspect in the life histories of environmentalists such as Anthony, and the environmentalists interviewed by Milton (2002), is not simply that their childhood knowledge of "Nature" led them in later years to "love Nature" and to express their love by wanting to preserve it. The key is that in these childhood experiences, the knowledge and love of "nature" to which Milton (*ibid*) refers was engendered especially during times of leisure (or holidays) as opposed to periods of work or school, which were not spent in places associated with "Nature". A comparison with Greek island farmers reinforces this point. The local farmers of Zakynthos also love nature (Theodossopoulos 1997). Yet they enact this love not through conservation but through "developing" the land, in other words through making the land fertile for human life (*ibid*). The Greek farmers' knowledge, and their consequent love of "Nature", are not engendered through periods of leisure, but on the contrary, through the work that sustains their lives. What makes the knowledge and love of "Nature" of the conservationists particular is that it is engendered in spheres of life separated from generative practises, as is employment. That is why there is a subtle preoccupation with apparent stasis: with conservation. During protests in Malta in 2008 against the development of a yacht marina in a place called Hondoq ir-Rummien, Gozo, one of the main arguments raised by environmentalists, including FoE activists, was the loss of summer destination they had enjoyed since childhood.[21] Conservationism, laced perhaps with nostalgia, attempts to preserve these memories (Milton 2002; Argyrou 1997, 2005). In conservationist approaches, the understanding of nature is highly shaped by experiences of rest, as opposed to periods of work. Places not densely peopled imply rest, rest equals stasis, stasis underpins a drive to conserve and preserve. Although, as Latour (2003) argues, in actual fact the conservation of nature involves more intervention than the notion of 'preservation' implies. The ideal upheld in these cases is that of retaining the "original", unpeopled nature (Gatt 2001).

Yet as Anthony's environmentalism and that of most others of the FoE core shows, even if a person's love for a domain called nature has its origins in an ideal of leisure, and even if that domain was originally defined as excluding humans, this does not mean that understandings are unchanging. Over the course of his life and experiences, Anthony shifted his understanding from an equation of nature with non-human to a concept of "environment" that includes humans and human problems such as injustice. Emanuela's shift, on the other hand, was from a concern with social issues to the inclusion of the environment in these issues. Nor do 'conservationist' origins restrict activists' ways of expressing their love of nature. Different experiences within and beyond childhood holidays have allowed FoE activists not only to have different understandings of nature but also to shift their actions and their activism.

Life Politics and Reflexive Modernisation

Giddens (2008, 2007) argues that within reflexive modernity, radical doubt replaces formulaic truth as the basis for social interaction. Formulaic truth generated authority and cohesion in what he terms "traditional" societies as well as early modernity (Giddens 2007). As a result of the radical doubt of reflexive modernity, and this loss of traditional truths, both Giddens (2008, 2007) and Beck (2007) argue that the process of individualisation in late modernity is accelerated; individuals are empowered to choose their own lifestyles, rather than having these dictated by traditional authorities, but simultaneously they experience anxiety caused by ontological insecurities. Giddens (2009, 2008, 2007) also argues that institutions appear to control more and more of people's public activities, which leaves individuals with the ability to control little more than their own private lives. Ironically, in his view, only private space is left to the individual to express political positions. This, he argues, leads to the politicisation of the self and what he calls "life politics".

Giddens (2008, 2007) maintains that having a "lifestyle choice" is a concept that is only possible within the framework of reflexive modernity, brought about by globalisation. Within the ambit of traditional authority, the choice of having a different lifestyle was inconceivable not only because of the sway of authority, but also for a lack of available alternatives. Now, he argues, we have no choice but to be aware of alternative lifestyles, and influences from around the globe. On the one hand, Giddens's (*ibid*) typification of life politics seems to mirror the FoE activists' way of being activists—from the notion of activism, which itself assumes the individual, to the way all their personal choices and actions are monitored and understood as part of their political environmentalism. The focus on planning for flexibility also seems to resonate with Beck's (1992) definition of risk society. In such a society, it is expected that imponderables such as future opportunities and risk can be calculated and accounted for in the present. FoE activists' planning for flexibility seems to reflect the need for constant change that Marx and Engels impute to 'bourgeois ideology' (quoted in Beck 2007). In addition, Wright (1994) finds that flexibility is one of four root metaphors described by organisational studies for capitalist organisations. One only has to think of flextime.[22]

Alternatively, we could look at planning for flexibility through Giddens' understanding of the individual within a risk society. According to Giddens (2008: 28), for the individual 'living in a "risk society" means living with a calculative attitude to the open possibilities of action, positive and negative, with which, as individuals and globally, we are confronted in a continuous way'. In this light, planning for flexibility could be understood as a response to risky choices that the activists feel they have to face. Put another way, planning for flexibility can be understood as a way to manage risk on the understanding that things are constantly subject to change. However, understanding FoE activists' planning for flexibility, according to Giddens's reflexive modernity, would lead us to imagine that their daily practises are common to anyone living in situations permeated by the 'late modernity'. It would be redundant to talk of a particular FoE "technology of the self" or discipline because in Giddens's reflexive modernity we have no choice but to shape our own lives. Nancy Fraser (2003) also wonders whether Foucault's notion of governmentality is still relevant today. Fraser (*ibid*) argues that Foucault's notion of governmentality was developed in relation to what she calls Fordist discipline, where normalisation was the ideal and the practise. In her view, current developments in the world's political economy that emphasise *de*-regulation and flexibility mean that Foucault's notion of discipline needs to be adapted for post-fordist situations. Although it was not my intention, I now suggest that what I have described in FoE activists' discipline is in some ways an adaptation of Foucault's discipline to situations in which flexibility, not normalization, are root metaphors.

I now also find that another ethnographic study of green lifestyles, by Shepherd (2002) in Australia, also describes environmentalists' rigorous practises in terms of reflexive 'discipline'. Shepherd uses Weber's characterization of rationally active ascetics in order to frame an understanding of environmentalist lifestyle choices within so-called late modernity.[23] She also finds Foucault's "discipline of the self" in Hellenistic and Roman periods, as well as Calhoun's (1995) history of new social movements in the early nineteenth century, and uses these findings to argue that the politicization of everyday life, especially personal life choices, self-exemplification and the vigorous application of virtue can be found at different times in human history. For this reason, she argues, the reflexivity characteristic of the environmental activists she worked with is not a unique feature of late modernity, as Beck (1992) and Giddens (2008, 2007) maintain, though the framing of the 'environment' as the benchmark against which to regulate and monitor their actions may be (Shepherd 2002: 138).

Were the reflexive wayfaring of FoE activists a common trait of reflexive modernity, then all persons within it would be equally disciplined and equally environmentalist. Not only are we not all environmentalists in the same way as FoE activists are, but even within their own families, such as the previous example of Lili, activists recognise a difference in how reflexive their siblings are. In addition, in writing and revising this chapter, I have started to monitor my own actions and I have realised that my personal environmentalist discipline has relaxed. I do not reflect on what and where to buy things as much as I used to when I was doing fieldwork and therefore spending a lot more time with FoE activists. The same goes for using the heating, water usage, planning my routes around the city, although for health reasons, I prefer my bicycle to the bus or car. In retrospect, I realise that I have become less disciplined and less reflexive. Were the activists' way of life not a technology of the self in a particularly disciplined manner, but only a common aspect of reflexive selfhood in late modernity, I would not have been able to become less disciplined or less reflexive.[24]

Finally, the differences in what is considered environmentalism, activism and personhood found among FoE activists are of vital importance when considering the generalisations of social theory about globalisation and environmentalism. Developments in transport and communication technology and mass media, as well as commodity and financial markets spreading around the world and creating paths of connection between previously unrelated locations, have been associated with the so-called process of globalisation (Inda and Rosaldo 2002; Eriksen 2003). Social theorists such as Giddens (2008, 2007) and Harvey (2000) argue that increased globalisation will lead to increased homogenisation. Giddens (2007) argues that globalization is leading to the 'evacuation of tradition', where any local meaning or truths will be, and increasingly already are, replaced by increased individualization, ontological insecurity and heightened reflexivity brought on by 'radical doubt'. Even when "local" traditions are upheld, he argues, they will not hold formulaic truth but will be recognised as choices of one tradition over other equally valid ones (Giddens 2007: 100).

If we consider Giddens's (2007: 85) statement that 'local tradition is increasingly evacuated and replaced by global influences', the question that arises is where do those influences come from? In Giddens's writings, they come from the processes of modernity. Historically, these processes derive from Euro-American bourgeois lifeworlds (Beck 2007). With this simple point in mind, the question of homogenisation looks more a spreading of Euro-American 'bourgeois' ontology. Anthropologists have amply shown how this homogeneity and decentring of knowledge is not coming about in the way Giddens predicted (Inda and Rosaldo 2002; Eriksen 2003; Clough and Mitchell 2001; Tsing 2005). They have also shown that there is multivocality even without globalisation (Marcus 1998; Geertz 1983; Lund 2005). In addition, I have shown how FoE activists come from different socio-economic backgrounds. The activists explicitly concern themselves with the possibility of imposing their own interests and understanding on others, or consider their privileged position to mean that they are required to work for change in the institutions and contexts to which they have access.

On the one hand, the welcoming of different forms of environmentalism, and the ideological embracing of diversity, seem to conform with Giddens' (2007) description of tradition and politics in late modernity; that emptied of formulaic truth, traditions will allow less violent politics. On the other hand, plurality does not seem to entail evacuation. Diversity continues to thrive. Rather than losing 'truthfulness' or 'meaningfulness', FoE activists learn different 'local' understandings, knowledge and histories from others from different countries, different socio-economic backgrounds, different political situations, as well as from non-humans, incorporating them to different degrees.

Notes

1 Many groups from Central and South America have for years struggled against the notion of individual rights, in favour of the recognition of community rights—they struggle against the appropriation of land by individuals, which, they argue, belongs to the community and therefore cannot be exploited by single individuals, and they have long-held, specific, academically based objections to private property. However, in conversations at the FoEI meetings, the most vocal supporters of community rights, also share, and are interested in, biographical information.

2 The emphasis of FoEI on making space for diversity, even of diverse understandings of environmentalism, without compromising a sense of hanging together or belonging, is also an ideological hope. The drive towards belonging not based on homogeneity is a political aim, in opposition to the homogeneity of categories expected in modern ideologies (Ingold 1993; Mol and Law 1994). See Chapter 6 on how FoEI reconciles these two different ways of belonging and Chapter 9 on how the latter understanding of belonging is also a political goal.

3 I am indebted to Tim Ingold for this term. Any mistakes in the way I have interpreted it are entirely my own.

4 'Persons' in my discussion refer to human and non-human persons, here I am following Bird-David's (1999) discussion of 'response-relatedness'.

5 The notion of co-respondence builds on the anthropological literature on the negotiated and contested character of culture (Cohen 2000; Loizos and Paptaxiarchis 1991), but develops beyond it by not requiring any idiom of 'culture', which implicitly retains notions of shared, bounded templates (Ingold 1992, 1996). See Brubaker and Cooper (2000) for a similar argument but in relation to the notion 'identity'.

6 A journalist, Teresa Urban, author of *Missão quasi impossível*, had interviewed Magda Renner as the founder of NAT and it was reported to me that a sociology PhD student had similarly interviewed a number of people in the office for his thesis.

 7 I am indebted to Tom Yarrow for the structure of this chapter—following his work with Ghanaian 'activists'. I adopt his approach in exploring in detail the life histories of a single person, respecting the individual life as understood by FoE activists.

 8 During the life history interview Anthony, as most Maltese speakers do, code switches between English and Maltese. To facilitate reading, I have removed most of the code switching, translating almost everything into English, but I have not entirely altered Anthony's sentence construction, which is characteristic of Maltese English. In a couple of instances, Anthony spoke Maltese for more than one or two words. In these instances, I have retained the Maltese due to the idiomatic reasons that led him to switch more entirely to Maltese.

 9 Alternattiva Demokratica, known by its acronym AD, is the Green Party in Malta.

10 In Malta, the description that something is 'political' carries different connotations. As Anne Zammit explained during her life history interview, "possibly due to the strong tradition of patronage in Malta, it is easily assumed that cooperation [with political parties] will lead to a long-term patron-client relationship." This was at a time when any disagreement with state policies, practises or personalities was considered political, by contrast to claiming disagreement on the basis of technical or moral reasons, which have an underlying claim to serve universal interests (Tsing 2005). During the 1970s through to the 1990s in Malta, the 'political', in the sense Anthony uses about his father, was largely associated with starkly pro or anti-state positions, and the 'political' remains largely understood in terms of party politics (Jon Mitchell 2002).

11 See Boissevain and Theuma (1998) and Boissevain and Gatt (2011), for details on the Verdala golf course campaign.

12 For these practises and experiments, see Chapters 6 and 8.

13 In their unintended lack of a central code and in their concomitant celebration of diversity, the discipline of FoE activists closely resembles the discipline of Quakers (Dandelion and Collins 2008; Dyck 2008; Collins 2008). The Quakers are also called the Religious Society of Friends (RSF), which increases the similarity and points to their preference for the sorts of affinal associations that Haraway (1991) talks about. In such affinal associations, or associations of affinity rather than consanguinity, association is not based on likeness; difference does not block belonging. Unlike FoEI, however, the RSF is a much older organisation, and its disciplined practise has been codified to a large extent. This may or may not happen in FoEI, although, as I discuss in Chapters 6 and 9, the growth of the Federation and the events that led to the SVPP seem to be bringing with them more codification of how discussions are held.

14 See for instance, Robin Eckersley 1989, Robin Grove-White 1993; Latour 2003; Anderson and Berglund 2003; Theodossopoulos 1997; Argyrou 1997, 2005; Milton 2002; Strang 1997.

15 See for example, Argyrou 1997; Croll and Parkin 1992; Theodossopoulos 1997; Anderson and Berglund 2003; Cristina Adams 2003; Tsing 2005; Strang 1997, 2004; Fortun 2001.

16 See for instance, Berglund 1998; Timothy Choy 2005; Gatt 2001; Boissevain and Theuma 1998; Boissevain and Gatt 2011, contributors to Milton 1993.

17 See for instance, previously mentioned works by Anderson and Berglund (2003), Little (1999), Ferguson (1990), Latour (2003), Fischer (1997) and Adams (2003). I am indebted to Tom Yarrow (2008, 2006) for this insight through his work on development NGOs in Ghana.

18 I briefly touch on the understandings and practises of service and sacrifice in Anthony's life history. It is a theme that emerged from most of the life history narratives, but due to limited space, I will not be able to expand on this here.

19 Indeed, the notion of learning from other people's experiences without necessarily having to go through them oneself in person is a currently explicit aim of FoEI. In the annual report in 2009—after the Swaziland meeting in 2008 where these points were discussed—one of the means adopted to implement the organisation's strategic plan was that FoEI was considered a 'learning organisation' — http://members.foei.org/en/resources/publications/annual-report/annual-report-2009/our-strategic-plan/guiding-the-implementation-of-the-strategic-plan/learning-within-friends-of-the-earth-international (accessed on May 13th 2010)

20 Tree plantations were being called 'green deserts' by their project partners in the other southeastern state of Brazil, Santo Espirito.

21 The comments of people remembering Hondoq from their childhood were mainly part of discussions and conversations during protests held in 2008. However, similar comments can be found on the website dedicated to protecting this coastal area, see the following web pages www.ipetitions.com/petition/hondoq/signatures.html (accessed November 2010).

22 Another resonance with Giddens' reflexive modernity is that activists do experience and refer to influences from distant parts of the world as part of daily lives. However, Giddens (2007: 96) characterises these influences as the predominance of absence over presence. In Chapter 7, I argue how these far away influences, which Giddens (2009, 2008, 2007) calls absences, are concrete and material presences.

23 Although Shepherd's use of Weber's 'ascetics' works well in her argument, I prefer not to refer to activists as 'ascetics', nor to use Weber's work, for two reasons. First, the central place of rationality in his argument is difficult given that FoE activists do not separate emotions from belief. In addition, see Milton (2002) on how interest is also an emotion. Second, 'asceticism' denotes a particularly religious take, and environmentalists are often characterised through the lens of religion (Douglas and Wildavsky 1983). Berglund (1998) finds the consequences of this characterisation problematic, as it positions the work of the activists as entirely about producing and reproducing the group itself, making no space for the non-human environment.

24 Collins (2008) argues that discipline can replace the notion of habitus and of practise, because it includes intentionality and covers the work done by the two latter terms. While I agree with his critique of habitus, I disagree with the view that 'practice' can be replaced by 'discipline'. That would be to dilute the specificity of what a discipline is. Although discipline can be understood within the terms of practise, to be disciplined is to refine one's practise, not only by definition but also in the terms of the Quakers with whom Collins worked, who want to *improve* their way of living in the world, and in terms of FoE activists who not only want to improve their own way of living but also to influence the regulation of others' actions through their campaigns. I have also mentioned how one's personal practises may become less disciplined when not involved in the same co-responding environment as other activists. But even when less disciplined, my daily life can still be understood in terms of practise. Therefore, I would argue that discipline refers to more specific situations than practise.

6 Rhythms of Globality
Developing a Sense of Belonging to FoEI

Activists generally first join a national member group and only gradually learn about the International Federation. Coming from their national member groups, activists are transformed by participating in FoEI international meetings. At these memorable events, the creation of a sense of belonging to FoEI, specifically to a federation that imagines having global-wide presence, is vividly apparent. A sense of the out-of-the-ordinary prevails. A FoEI international meeting is a 'bubble' of intense affective time that has altering effects on activists.

In both FoE Malta and FoE Brazil, activists who have been to the international meetings, especially those who have been several times, often talk about these often back home. Even the members of the IS, who have constant contact with the transnational activities of FoEI, are similarly excited about the international meetings. Anthony, FoE Malta's coordinator, recalls Julian telling him about the wild FoEI parties and about 'Annamaria', an experienced representative of FoE Italy who regularly attends the international meetings. Julian described her as large in size, voice and presence. Then when Anthony met her, he confirmed all those things, but also added that she was very patient, showing him around and explaining things. He recounted all these details to me as a form of preparation as I was planning my participating in FoEI international meetings. When I, then, first saw Annamaria at a FoEI meeting in Abuja, she was standing up to talk in a plenary meeting. No wonder she has an expansive presence, I thought, apart from anything else no other activist stood up to talk. When we were introduced later, I found the descriptions Anthony had drawn of her were very lifelike. She was knowledgeable about FoEI history and patient with her knowledge. But I also found that she loves to play cards in the evenings instead of joining the parties at the meetings, and that she was tired of politics and looked forward to a peaceful retirement. Although experienced activists talk about the FoEI meetings and the people they meet, these descriptions remain characters in stories. Vicariously imagined these personas remain nonetheless flat in comparison to the vividness and altering quality of the actual experiences of the international meetings.

In this chapter, I explore the different ways that FoE International is portrayed and most importantly experienced as a 'global' grassroots federation.

The question that follows is broad: how do activists come to experience FoEI as 'real', when they will never meet most of the persons that together compose it? Benedict Anderson's question about belonging to a nation-state morphs in particular ways when belonging is characterised by global scope. Participation in the meetings is a significantly transformative experience, more than the stories told suggest. The specific character of the changes in identification and sense of belonging that activists undergo in the international meetings is a pivotal refraction in the spectrum of FoEI's global imaginary of its environmental activism.

FoEI International Meetings

Every two years FoEI holds a General Meeting (Biennial General Meeting— BGM). Each BGM is hosted by a different FoEI member group each time. The General Meeting (GM) is the ultimate decision-making body in FoEI. During these meetings, an Executive Committee (ExCom) is elected to run the day-to-day decisions of the Federation. The ExCom is composed of activists from different FoEI member groups with the aim of obtaining 'regional balance'. A typical agenda of a BGM will include reports from the ExCom and the programme coordinators; issues about continuing or changing programmes, campaigns or policy; a financial report, amendments to the constitution and new membership applications.

Apart from the BGM, there are other types of meetings understood as part of the process of constituting the Federation as a whole. A large international meeting is also organised in the intervening year between BGMs, informally called "inter-BGMs". Each FoEI region has its own annual meeting. The different FoEI campaigns are organised into programmes and each programme has a coordination group. Depending on the programme, the coordination teams formally meet up to three or four times a year. The member groups often participate in joint projects, sometimes these are exclusively FoEI projects.[1] At other times, such projects include other, non-FoE, organisations. As part of these projects, activists may meet face-to-face for planning or 'actions'. Apart from face-to-face meetings, activists work together on FoEI issues by telephone, telephone conferences, individual email and email groups. Even though most of FoEI's daily work is now done through ICTs, considerable amounts of energy and money are still invested in organising regular meets. The face-to-face meetings of the FoEI Federation are greatly valued in the eyes of core and intermediate activists. The GMs are more important than the campaign meetings for 'solidarity', whereas the campaign meetings are essential for the detailed policy and campaign work of FoEI, at least since 2004 and the beginning of the Strategic Vision and Planning Process (SVPP). In this chapter, I focus on the BGMs, inter-BGMs and the AGMs of one regional grouping: FoE Europe.

Applicant member groups spend two years as associate members, with no voting rights, until they are voted in as full members. The international meetings also introduce activists who may be veterans in their national member groups but are new to FoEI and the workings that constitute it. In many cases, the same activist annually represents and participates in the GMs. Some member groups can only participate in FoEI by attending these meetings, in these cases the Secretariat and ExCom affectionately call these regulars 'the faces of the FoE groups'. FoE groups are bodies, political bodies, but also organic bodies, that have faces as well as organs.

For many veteran activists, the international meetings also carry an echo of an origin myth of FoEI. In their historical accounts, FoEI was founded in "the annual meetings of environmentalists" (Brower in an interview with Jonathon Porrit 1991). For some, though by no means for all, the international meetings are implicit enactments of the original FoEI meeting.

Abuja BGM 2006

In 2006 Environmental Rights Action (ERA), the Nigerian FoEI member, hosted the BGM. The meeting lasted for seven days. GMs are generally divided into two broad phases: the pre-conference, which is open to a broader public, and the GM itself, which discusses issues considered internal to FoEI. The BGM in Nigeria, for instance, approved a ten-year strategic map for FoEI, and voted in eight new full member groups to FoEI.

On the final day of this meeting, once the delegates had discussed the individual membership applications and voted in turn for the inclusion or exclusion of the groups in question to full FoEI membership, the chairwoman of FoEI invited the representative for each new member group into the plenary room. She explained the outcome of the voting to each individual representative and welcomed them in as full 'FoEI members'. When all the successful applicant group representatives had been called in, she proclaimed, "Welcome to the Foe-I family!" All the other delegates spontaneously rose to their feet and clapped heartily. As the clapping went on, the new delegates took their seats around the table. Waiting for them there was a placard with the name of their country: Guatemala, Sierra Leone, Ireland, Belgium Flanders, Bolivia, Swaziland, Nepal and Bangladesh.

Imagining FoEI

Before they have participated in the international meetings, the activists I worked with tended to imagine FoEI in a particular way. I also shared something of this variable imaginary as an activist with FoE Malta between 2003 and 2005. Newcomers talk about FoEI as an almost opaque, sometimes solid, thing. For Julia, an activist in the FoEI Secretariat, the process of getting to know more and more people in FoEI felt like open, light space clearing away dark, cloud-like obscurity. Anthony described the process as getting to know different pieces of a jigsaw puzzle, first through the stories Julian had told him, then by meeting people at meetings, and slowly becoming more familiar with more and more people, as well as with the way things are done. He became more familiar with the stories, topics and events that regular participants talked about, making it easier for him to join in conversations. However, before going to any FoEI meetings, I, like Anthony, had read about the aims of FoEI and the different campaigns of FoEI. I even had to present these to the thirteen new colleagues from around Europe in the first meeting of the EU project that I was employed by FoE Malta to coordinate locally. At the time, I had what now seems like a two-dimensional understanding of FoEI, based on what I had read. At that meeting, held in Malta, I also met Elena, the Cypriot coordinator for this EU project. She was the long-standing coordinator of FoE Cyprus, who had worked with FoE Malta on various MEDNET (FoEI Mediterranean Network) projects. She had come to Malta before, and had met a number of other FoE Malta activists at FoEI, FoEE and MEDNET meetings. Julian organised a meal while she was in Malta, and many FoE activists from earlier times, who no longer participated in FoE Malta activities, came along. It was a reunion of old friends, although, as a newcomer, I could only see that as an outsider. But I could recognise the warmth in their manner towards each other. The friendship was discernible from the similar way Julian, Anthony and Raffaella, with whom I had already become friends, behaved towards me.

Even with FoE Brazil, other FoE activists sometimes come to visit. For a long period the World Social Forum (WSF) was held in Porto Alegre every second year. For six years running, FoE Brazil helped organise the FoEI events held at the WSF. This included finding venues for events, setting up an office for the Secretariat while they were there, organising tents for other FoE groups, organising meals and a couple of social events in Amy's home to bring together the FoE activists after the WSF. Andrea, who speaks English and German fluently and who had not only attended FoEI meetings but also spent a week with FoEI activists at the protests against the WTO in Prague in 2001, talks of those meetings as having 'friends over'. However, others who are also long-standing members of FoE Brazil—including Amy, who joined FoE Brazil before Andrea, but who had not been to international meetings—do not talk about these events in the same warm terms. Not being able to speak English, Amy did not feel personally close to the other FoE activists who visited Porto Alegre every other year for the WSF. Noemi remembers being the '*fac totum*'. She recalls activists from other FoE groups searching for 'Noemi! Noemi!' But since her role was to help them with her 'local knowledge', she did not have time or the opportunity to get to know them. She went home every evening, while the visiting activists shared all their meals, their working days, their evening socialising and their tents at night.

Apart from hearing stories and meeting with activists visiting from other FoE groups, newcomers to FoEI are presented with documents. In FoE Malta I was given the FoEI mission and a list of FoEI campaigns to read and then to present. In FoE Brazil Andrea makes regular, sustained efforts to situate their 'national' work within the goals, values and mission of FoEI.[2] Newcomers first come into contact with these and other documents that portray FoEI as an entity that existed before they joined, might be changed by their participation, but will carry on existing when they leave. The existence of FoEI as some*thing* to get to know is conveyed through the FoEI website, especially those parts that explicitly claim the existence of a thing called 'Friends of the Earth International'. Take for instance the way 'FoEI' is described: there are maps of 'FoEI' regions, pages dedicated to the work of 'FoEI', and pages suggesting how visitors to the page can contribute to 'FoEI'.[3] The same can be said of the annual reports and other such 'public' documents in which 'FoEI' is portrayed as a stable entity.

These representations are often more strategic than anything else. In the case of both FoE Malta and FoE Brazil, during their planning and discussion meetings with other ENGOs, the activists' identification with their FoE group is minimal. In these cases, what comes to the fore is their different skills and knowledge. For example, at a meeting of the Nature Group in Malta, a coalition of most of the Maltese ENGOs, the names of the different groups that activists belong to are rarely mentioned. The names of the organisations then reappear in press releases.[4] The intended audiences for speeches and documents, in some cases within FoEI's 'environments of legitimation' (Lister 2003), influence to what degree the relatedness between activists and the existence of a thing, 'FoEI', 'the Federation', 'the network', is talked about at all.[5] For instance, Stine, the finance manager of the IS, explained that the central purpose of FoEI's annual reports is to give FoEI funders an account of what their money was spent on, so that they will carry on funding FoEI. For these annual reports, FoEI member groups contribute descriptions of work during that year they consider successful: a golf-course planning application turned down after ten years of campaigning and negotiating (FoE Malta 2005); a priest mobilising a group of young people to plant organic vegetable plots (FoE Italy 2006); FoE Brazil organises a large conference to discuss the intensive planting of eucalyptus trees in the Pampa ecosystem (2007). None of these specific examples was intended as part of FoE *International*—there is a separate section of the annual report for specifically FoEI actions, however, like individual activists the actions of national FoE groups is understood to constitute FoEI.[6]

In FoE Malta, at the time I worked there, Anthony and I were the only two people who were actively interested in FoEI, who attended the international meetings or who had access to emails from the international campaign groups or the Secretariat. In FoE Brazil, where only two activists speak English well enough (in their view) to attend the international meetings, and although there is collaboration with other ATALC members, FoEI-specific work is limited. In other words, the image of a unified, self-aware FoEI portrayed in the annual reports is more a product of the editor's abilities than of the activists' actual orientation to FoE *International*.[7] The key point is that to newcomers or those who have not participated in FoEI international meetings, these documents portray a stable FoEI. More importantly, FoEI and other FoE groups are portrayed primarily as a federation of *national* member groups.

As only one FoE member group is permitted per nation-state, the activists participating in the international meetings (and in FoEI in general) do so as representatives of their *national* member group. The national membership structure directly affects FoEI practices. Most saliently, it affects what the activists are expected to contribute to these meetings and their other FoEI work. The national membership structure also directs where meetings are held, where FoEI activists work, where they get their money, how FoEI gains legitimacy as a widely 'representative' Federation in international political negotiations, where they direct their campaigns, and the internal politics of the Federation.

When pushed to explain the continued maintenance of the membership structure, some activists argued that it is in the interests of worldwide representation to organise this way. Many stressed that since nation-states are still the most influential actors in public international political processes, it still makes sense as a model. Others prefer to contrast FoEI's structure with international organisations such as AAVAZ.org, with which they sometimes cooperate. The latter organisation is often criticised for imposing a narrow idea of environmentalism on 'local communities', because their organisational structure is supposedly 'global', but is, in FoE activists' critiques, in reality composed of a maximum of thirty people, all from 'Northern' countries. Greenpeace and WWF receive the same criticism, that they are top-down and centralised. FoEI's membership structure, based on national member groups, is considered by most activists I worked with to be the best of the currently available solutions to ensure 'widespread and equitable environmental stewardship'.

Other FoE groups and their activists are presented to newcomers mainly through the idiom of nationality. To describe Annamaria's loudness to me, before I had met her, Anthony resorted to a stereotype of national character, referring to her as 'typically Italian'. He described the South American activists with similar stereotypes, noting their propensity to give long speeches. Considering his rich relationships with Annamaria and a number of activists from South America, I am convinced that his use of such national stereotypes is only meant to facilitate description. These stereotypical descriptions of specific activists through their nationality, and of the member groups as primarily defined by their nationality, portray FoEI to newcomers through a particular lens. FoEI is portrayed as hanging together in the same way as does the UN: as an assembly of representatives of different nation-states. They are representative because the logic of nation-state assumes internal homogeneity, and they cohere as a mini-version of the mosaic of nation-states. This, in turn, projects back onto the national member groups, who are then understood to cohere internally in the same way that nations do.

In the logic of nation-states, belonging arises from members sharing a set of attributes: a shared language, history, culture, spirit and geographical territory (Anderson 2006 [1983]). Creating distinctive groupings or categories according to internal homogeneity is not limited to nations. Many forms of grouping within the logic of modernity posit cohesion on the basis of puta-tive internal similarity (Mol and Law 1994; Brubaker and Cooper 2000).[8] The flip side is that those who do not share the same attributes are consid-ered outsiders. Ingold (1993) calls this identity-by-contrast, and since it is based on the supposed sharing of attributes, he refers to it as *attributional* identity. Within this logic difference becomes 'disjuncture' and a barrier to mutual belonging (*ibid*). For those activists engaged in FoEI activities, expe-rienced international meeting participants, and those regularly in touch with other FoEI activists, their national (attributional) identity is not diluted. And yet, their national identities do not reduce their sense of mutual belonging within FoEI.

Unlike attributional or categorical identity, where cohesion arises from attributes that already have to be shared to start with, immersion in joint action develops *relational* identity (Ingold 1993). This is akin to what Jack-son (1983: 341) refers to as a 'field of practical activity' which provides "[consonance] with the experience of those among whom one has lived". Relational identification therefore is how one positions oneself in relation others (Brubaker and Cooper 2000). Here, 'difference' is not a barrier to mutual belonging but the *basis* upon which persons engage with each other. Difference is a function of every participant's positioning 'within a continu-ous network or field' (Ingold 1993: 229). In relational identification, differ-ence is *constitutive* of a sense of belonging.

Ingold (*ibid*) argues that people develop a sense of attributional iden-tity when their relational ties are threatened or lost. Barth (2000: 32–33) offers a poignant illustration of the shift from relational to attributional identity. He relates the work of Bringa (1995) before and during the con-flict in Bosnia. Bringa shows how the members of a village are related to each other primarily on the basis of neighbourliness, but also love, friendship, work, exchange, affinity, kinship and only in certain circum-stances by religious congregation or ethnicity. However, as rumours and subsequently experience of inter-ethnic violence increased, congregation and ethnicity became more salient and the other types of interactions were slowly cut off. The inhabitants of this village came to experience an actual breach in their relations, leading to the consolidation of a sense of ethnic identity which, as with national identity, was couched in attribu-tional terms.

With FoEI activists the opposite seems to occur. The activists from dif-ferent national member groups start out as *not* having personal relational ties with each other or as regards to FoEI. They do not generally know or have friendships with other activists outside of their national member groups. In the documents and stories they are told, FoEI seems to hang together based on a collection of attributionally based identities (corre-sponding to national member groups see Figure 6.1). However, whereas the initial understanding of activists and their learning about FoEI is primarily through the national membership structure, which is attributional, the FoEI

Figure 6.1 Cake at the FoEI BGM in Nigeria showing the national flags of the FoE member-groups.

international meetings provide the ground and a particular atmosphere that engender relational identification. Simultaneously, the international meetings also create boundaries vis-à-vis non-FoEI activists.

Dividing FoEI from Non-FoEI

Host groups have always attempted to use the presence of so many international delegates at international meetings to support an issue that it is working on. While this is aimed at showing and building solidarity, the shift from pre-conference to FoEI BGM creates boundaries. The Abuja pre-conference was entitled 'Energy Sovereignty', hinging on the long-standing oil-induced crisis in the Niger Delta. Apart from all the FoEI participants and observing academics, those invited and present included a number of journalists, government representatives and inhabitants from the Niger Delta, who were part of the struggle against the oil companies and against the government. The members of the local community, as they are called by FoE activists, have an ambivalent position in relation to FoEI.[9] But the gap between FoEI activists on the one hand, and journalists and government representatives on the other, was blatant.

The pre-conference was set up in the form of a panel of speakers, seated behind a long table elevated on a platform, using microphones and clearly introduced as representatives either of the Nigerian government, the local communities or a particular FoE group, before an audience arranged in rows of chairs was facing forward. The pre-conference is a public event, whereas the BGM is a less public event that only FoEI representatives are allowed to attend, and in the case of Abuja the invited scholars.[10] The FoEI business meetings are closed even to invited guests. In addition to the fact that the activists travel to get to the conference, the shift from public pre-conference to 'internal' BGM adds to the process of isolation from people not considered FoEI. This shows that the activists *do* erect boundaries, and identify themselves together against non-FoEI 'others'. Although FoEI is experienced as a global federation, here global is not all encompassing.

The activists, at least the core activists, experience a strong sense of belonging as they work together in FoEI. They also experience a sense of commonality, although again to different degrees depending on the activist and, importantly, commonality is not the primary motivation for maintaining their associations with each other. As we saw, they also create boundaries *at times*. The activists certainly feel that FoEI exists as something they belong to and that they constitute. The international meetings are the most prominent place where we find an explicit expression of a strong collective sense of belonging, a groupness (Brubaker and Cooper 2000). However, in the activists' experience of FoEI, these boundaries do not enclose a commonality that erases their differences. Further still, this sense of boundedness is not present all the time, even with the most partisan of FoEI activists. Their attributionally based identities are not simply replaced by relational belonging at the meetings. Even for experienced activists national identities remain important; you are either born Columbian or German, though by means not for all. The case of the FoEI secretariat members complicates this (see Chapter 8).

Another way of describing how belonging is generated in an environmental federation structured along national lines would be according to the model of 'unity in diversity' as in classical, Durkheimian *organic solidarity*. According to this model, the nations of the world would be linked in an immense division of labour, each responsible for looking after its particular segment of the Earth. Indeed, the ideal of the United Nations, that FoEI at one time mirrored, as well as the notion of regional ecosystems still strongly upheld in FoEI, conforms to this model. However, as I showed in Chapter 3, FoE activists simultaneously perceive the entire world as a seamless place, in which such divisions do not make sense. It seems that activists incorporate both national identities, based on the supposition of shared similarities with co-nationals, *and* FoEI belonging, grounded in diversity. Ingold has already given us a clue as to how this could work. To restate this particular point, he argues that a loss of personal relations, such as the loss of neighbourliness described by Bringa, results in an emphasis on attributional identification. However, in the case of many co-national FoE activists they share both attributional identity as well as an ongoing common ground of shared activities. In other words, sharing some aspects, or at least having the impression of sharing some attributes, such as membership in a particular nation-state, does not exclude relational identification where homogeneity is not the grounds for belonging.

Getting to Know FoEI

Experienced FoEI activists are considered those who have often participated in meetings, and are knowledgeable about FoEI in a rounded sense: they know many other activists, the stories of how campaigns developed and their content, the way things are done at meetings, and they keep in touch with and work on all these things in between meetings. These activists create representations of what FoEI is like for newcomers. The representations come in the form of the documents and stories that I described previously. FoEI's existence as a *thing* is also conveyed in certain turns of phrase. Experienced activists, primarily in each other's company, but also with newcomers at meetings, when speaking English, refer to Friends of the Earth International as FoEI (pronounced foe-I), or as 'the Federation' or 'the network'. They talk of FoEI as if 'it' existed enough for activists to talk about 'it', for governments, foundations or other institutions to fund 'it' and for 'it' to have an effect in the politics of environmental justice. It is also common for activists to use the first-person plural when talking to each other. They talk about "who *we* are" or "our goals" in the same way that the website has a section "about *us*"—as though the we/our/us were uncontentious. These representations are part of what gives rise to a FoEI that exists enough to hang together. To newcomers, as well as non-FoE activists, the documents and stories create an imagined community, FoEI.

The strategic 'thingification' of FoEI that I describe does not necessarily imply that 'FoEI' is not *real* (Taussig 1992).[11] Strategic portrayals of FoEI as a thing are neither in opposition to nor reduce the significance of other non-strategic forms of relatedness or a more personally felt sense of belonging. In fact, these strategies *contribute* to the institutionalisation and crystallization of ephemeral relations into tangible things: an office; a website; publication of a document; a campaign group; the international meetings (as well as other FoEI projects, actions, staff exchanges and so on). The activists' strategic essentialisation is part of what generates the coming-into-being of FoEI, in other words how it exists enough for it to hang together at all. The creative, strategic process of imagining is accepted not only by activists but also by funding bodies. By their very function, funding bodies acknowledge that by needing (and their giving) funds this 'thing', FoEI, needs support to carry on 'being'. Resonant with processual approaches in academia, experienced activists regard 'FoEI' as arising from the work of activists and FoE groups. The imagination, strategic portrayal and effort involved in bringing about these plans and dreams are part of that work.

Above, I described how newcomers talk about FoEI as an almost opaque, sometimes solid, thing. From the activists' narratives, the perception of a solid FoEI is the result of a newcomer's lack of familiarity. A similar argument is made by Jackson (1983). He suggests that the perception of a *system* of meanings is the consequence of outsider status: the result of a purely observational and non-participatory role. However, the impression of endurance, at least with FoEI, is not merely a belief or an illusion. With time, engagement and participation in FoEI, newcomers lose the sense that FoEI is *opaque*, but they do not lose the sense that 'FoEI' exists, and will continue to exist when they leave, and existed before they joined. 'FoEI' may come to seem more fragile, but even then, this perception of fragility often generates *more* commitment to maintaining 'FoEI'. For instance, Andrea (the coordinator of FoE Brazil) is disillusioned by FoEI work because people do not fulfil their commitments, yet she more than ever wants to keep FoE Brazil oriented towards '*foi*' (as the Brazilian activists call FoEI). During their annual strategic review and planning meeting in 2006, she led a discussion focused on aligning FoE Brazil's aims with the newly instituted FoEI mission, vision and values.

Another instance is in Roberta's plans. She is a campaign coordinator at the FoEI IS, and was employed in part to manage the conflicts in a number of FoEI campaigns. Roberta was painfully aware of the possibility of the mass withdrawal of South American FoE groups in the wake of the 2002 crisis with the then FoE Ecuador group. Though this possibility was allayed through the SVPP process, Roberta explained that she would like to develop FoEI solidarity by someday proposing members of the Secretariat based around the world in order to increase decentralisation. Roberta is articulately conscious of the instability of 'FoEI', and actively imagines possible ways to stabilise it.

Thus, the portrayals and imaginings of FoEI as a thing, whether explicit or implicit, are part of what continues to bring FoEI into being. Furthermore, what looks solid from the outside opens up from the inside, to look airy, translucent and even fragile. These imaginings lead to the crystallisation of FoEI around tangible things such as funds, offices and documents. A crystal is indeed a perfect example of something that looks from the outside like a solid substance, with hard, angular surfaces, but that—understood from the inside—reappears as a delicate *lattice*, held together by lines of force between its constituent molecules. So it is, too, with the FoEI.

In one way then FoEI is an imagined community in Anderson's terms (2006). FoEI members will never meet every other FoEI member in person; especially since who is considered a member and who not is also fuzzy. Print media (*ibid*) are also central to the constitution of this imagined FoEI.[12] Additionally, this 'FoEI' mobilises activists to work in its name. The activists' imaginings of FoEI are creative actions. These imaginings are then translated into tangible actions that, more or less according to plan, bring about 'FoEI'. Moreover, while FoEI activists will never meet all the other reported two million members, the imagining of FoEI *drives* the core activists to organise meetings with more and more other activists, whether face-to-face, or with ICTs. Therefore, together with the strategic thingification of FoEI, such creative and active imaginings also contribute to bringing about other relations that are not only imagined but also experienced. And this is also a way to understand how there need not be such a stark ontological distinction between imagination and 'real life' (see Ingold 2010), making space for imagination in ecological understanding of sociality.

It is most common that activists' first experience of FoEI is through imagining. Through the meetings many activists, but by no means all,[13] then go on to develop relational ties with other activists, with people that they are only engaging with in the first place because they are part of this imagined FoEI. They then develop these relations through the work they carry out together, whether at subsequent international meetings or by means of ICTs. The very real places generated through ICTs are the subject of Chapter 7. For now, I focus in more detail on what happens at the international meetings: what brings about the more personally felt sense of belonging to FoEI?

Bubbles: The Special Zone of International Meetings

At the time of my fieldwork, FoEI had seventy-one national member groups. Participants in the international meetings have to travel varying distances to get there. Around half the participants at every meeting have long-distance flights, and many need to plan their trip months in advance not only to book their transport, but mainly because of the complications of getting visas. Once the actual journey has begun, many activists meet up with other FoE activists well before reaching the venue of the international meeting. For the Abuja meeting, a group of South and Central American FoE representatives met in Buenos Aires airport to catch a transatlantic flight to Madrid. They spent a week touring Spain giving talks about the situation in Uruguay, Argentina, Peru, Guatemala, Colombia and Costa Rica. They then travelled together from Madrid to Frankfurt and there caught the plane to Lagos. Travelling from Scotland I also travelled to Frankfurt International Airport, where I met Anthony, coming from Malta. Anthony introduced me to three or four other activists whom he knew well. We sat together on the plane journey that lasted seven hours. A couple of times during the flight other FoE activists would saunter along the aisles looking for friends to chat with. According to Anthony there were at least fifteen other activists on the same flight.

During the flight, Anthony and I discussed many things, caught up with each other's news, but mainly talked about FoE happenings, gossip from FoE Malta, and stories about people we would be meeting in Abuja. We discussed our ideas about sustainability, and disagreed intensely about the implications of long-term cycles of biodiversity. More than once, Anthony took photos of the landscape we could see from the plane window, especially when we were flying over the deep red earth of the African landscapes that neither Anthony nor I had seen before. We noticed how the contours of the landscape looked flattened from this height, like the contours of maps. When we were still flying over the Mediterranean, familiar to us as rough seas, waves crashing on familiar rocks, or dead calm seas that in Maltese we say looks like *żejt* (oil), we marvelled, not for the first time, at how the 'white horse' crests of big waves looked perfectly immobile. Anthony and I were not the only ones to marvel at the view of the world seen from above, from a plane.

On the way to Swaziland, around a year later, Lili (FoEI Secretariat) spent a lot of the flight taking photos through the window, telling me how she likes to compare what things look like from the air and on the ground. In Chapter 2, I described how aerial photos are an important part of the watchdog work that Veronica used to do for FoE Brazil. More than once, she employed an aerial photographer, and went with him, to monitor the activities in a river valley that was part of one of their campaigns at the time. The doors of FoE Brazil's office, and similarly with FoE Malta and FoE IS, have various maps hung on them, which bear a strikingly resemblance to the way the ground looks in the aerial photographs, and the view from the plane windows (see Figure 6.2). In Chapter 2, I also described how Google Earth and Google satellite maps, that FoE Brazil activists began using regularly, have extended that ease of viewing the Earth from such dizzying heights. In the same way activists combine their FoEI belonging with their national identities, they also combine their imagination of the world as the environments they walk through and live in, with the object-like 'Earth' portrayed in globes, maps, Google technology and aerial photography. The two ways of perceiving the world, as a sphere and as a globe (Ingold 1996), coexist in their daily lives. What activists say about the Earth, and the environment, as something that in their images of a global world is solid, opaque and immobile, like the sea from the plane window, opens up as they get to know different places through the activists they meet at the international meetings. And similarly to FoEI, as they get to know more and more about the environmental struggles of activists from around the world, that Earth appears more fragile, in need of 'friends' to protect it and those inhabitants who ask for support. However, the image of the Earth as a whole, as a globe, does not disappear, just as FoEI as an entity does not disappear.

Once we arrive in the airport we meet others, having arrived on other flights gathering together to be transported in a bus to the venue of the international meeting. On the shuttle bus, activists already begin to ask each other where they are from, and where they had travelled from; getting to know other activists in the buses organised for us. The common ground of engagement as FoE activists begins slowly to be defined.

Figure 6.2 Photo taken from the plane on the way to the FoEI inter-BGM, Swaziland 2007.

Arrived at the Meeting, Abuja 2006

Around sixty activists and four academics participated in the BGM in Nigeria. All the activities of the meeting were held in a hotel, of which the FoEI delegates occupied nearly all the rooms. The exceptions were three outings, when everyone bundled into two buses to visit a market, Abuja centre and an impressive waterfall all at least half an hour's drive away from the hotel. In practice, this meant that participants spent most of the seven-day meeting within the same places in and around the hotel. The venues chosen for these meetings are usually relatively isolated. In Nigeria, in particular, the isolation from other human centres of activity was further exaggerated because the participants were strongly advised not to leave the hotel for security reasons. Two heavily armed soldiers at the gate emphasised the point. The meetings are held away from very populated areas as the organisers want to maximise the time spent together.

In 2002, when the BGM was held in a small town in Switzerland, in the evenings many participants dispersed to different pubs rather than spending the evenings together. Annamaria, who was then a member of the ExCom, explained to me that since then more isolated locations were preferred by the organisers. As will become relevant further on, in Swaziland the IS thought the hotel chosen was too luxurious, and this went into the organisation of the next meeting in Honduras, where the meeting was held in a basic corrugated-iron-roofed structure, and delegates lived in huts around the building.

Participants spent little or no time using computers due to the tightly packed agenda and lack of wireless Internet. Over and above the joint work during the formal meetings, they followed a more or less joint rhythm of waking, meals, work sessions, periods of rest, coffee breaks, cigarette breaks, evening entertainment or socialising, and retiring to their hotel rooms, also shared with another participant. During the meeting, participants spent most of their time focusing on tasks and activities *together*. There were rare moments of being alone or of not paying attention to what was occurring in the group.

For the most part, in activities on the formal agenda and in non-scheduled ones, people formed round shapes, either around a table, standing talking or dancing in a circle, or in a large oblong during plenary sessions. In these rounded shapes, the orientation of the participants' torsos and their attention was focused inward, creating joint rhythms. In the evening, many activists joined the dancing, also in larger or smaller circles of people. In the conversations, people took turns to talk, both formally with lists of speakers and informally during cigarette breaks or at the bar. This taking turns created a roving pulse: people turned their heads and slightly shifted their orientation to the next person who spoke, also shifting their orientation to different people if they were speaking, diffusing their attention among various listeners. The meetings are characterised by intense sociality and shared rhythms. The intensity comes from spending time in the same places with other participants, much more than in their daily work in their respective offices and where they would spend most of their time at their desk, on their computers. These conditions have led some of the participants to feel that these meetings are 'bubbles'—intense and inward looking, and detached from their daily lives and from the rest of the world.[14]

In a bubble-like way, time feels different to activists at these meetings. Approximately one year after the FoEE Estonia meeting at the FoEE AGM in Hungary, there were a number of people whom I had last met in Estonia. Laide (FoE Denmark) and Aurélie (FoE Spain), in separate instances, both exclaimed how strange it felt to meet again. To them and to myself it felt as though we had only parted yesterday; the time in between the meetings felt different, almost as though it did not exist, and the time associated with FoE meetings was almost continuous across different meetings. The participants feel that the meetings are different from their daily experiences because of the intense sociality and in their peculiar sense of time.

On the one hand, the 'special' time of the meetings, created by intense sociality and joint rhythms, makes the relations developed during these meetings valuable. Not because such gatherings lift people to a higher level of collective life (Durkheim 1973) which they cannot experience in their home countries, but simply because the change from daily routine makes them more memorable (Milton 2002: 65). In my own experience, after any of these meetings my surroundings felt acutely *un*populated. On the road, I often thought I saw or heard the voices of activists I had met at the meetings. Others have reported similar experiences on their return from meetings. Activists subsequently continue to invest in newly formed friendships through FoEI work relationships, or through imagining their belonging to FoEI. Importantly, rhythms help to create coherence among people. Rhythm is a variously patterned reiteration and, as such, it creates anticipation (Lincoln 2011). In this, the rhythms at the international meetings create the anticipation of working together as part of FoEI.

Brubaker and Cooper (2000: 17–19) define self-understanding as a tacit process of identification. Self-understanding incorporates a practical, Bourdieuesque 'disposition'; the implication for action in how one locates or identifies oneself. This is similar to the notion of orthopraxy (Carruthers 1998; Collins 2008) that works well in relation to FoEI. As with the Quakers that Collins (*ibid*) describes, FoEI is a largely creedless Federation that celebrates the diversity of its members' environmentalisms. Orthopraxy is a disciplined *way* of doing things, as opposed to adherence to the authority of canonical texts (Carruthers 1998; Collins 2008). At the FoEI meetings, this is referred to as 'process'. Newcomers often encounter for the first time the different processes of discussion (marketplaces, breakout groups, plenaries), attitudes towards listening (roving pulse, value given to different narratives of activism) and decision-making practices (inner/outer circle, collaborative consensus) which are currently highly valued in FoEI work. However, due to the grassroots and inclusive ideology of FoEI, the new activists also contribute, to an extent, to creating this orthopraxy.

Although Kaarina, the Finnish representative, had been president of FoE Finland for almost two years, Abuja was her first experience of a FoEI BGM. There had been a long discussion about FoE Palestine's membership application, which lasted almost half a day (and had been going on since 2004). The participants continued to discuss the issue in the coffee breaks and lunchbreaks. In the afternoon, Kaarina formally petitioned all the participants "not to take these discussions to the corridors". "If we have something to say we should say it in front of everyone, where everyone has a fair chance to respond", she explained. Although talking 'in the corridors' seemed to be what most people were doing, they agreed and from then on only non-controversial FoE matters only were openly discussed at break times. This orthopraxy can also be found in more tacit ways of doing things and in 'reading' each other's ways.

In the following section, I tease out what actions bring about collaborative consensus (as opposed to compromise consensus), and the recognition of particular persons' presence through their voices, ways of speaking and writing.

FoEE Meeting in Estonia

The first three days of the 2006 Annual General Meeting of FoEE in Estonia were dedicated to SVPP workshops. These workshops differ from the types of meetings held during the AGM because they are based upon breakout groups—smaller groups that work together on a topic, discuss the issues, prepare the results of their discussion and finally present these results to all the other participants. The shape and practice of work is different in these breakout groups from the shapes and practices in the plenary sessions.

In the plenary sessions, all the representatives of the European member groups sit at tables laid out in a large square, with the name of the country they represent on a paper label propped up in front of them—for voting procedures. During these sessions, participants speak to the whole group and the participants face the speaker while they listen. On the other hand, during the breakout groups, one may not necessarily be facing other group members while they speak.

One of the breakout groups that I participated in at the FoEE AGM in Estonia was aimed at imagining a ten-year strategy for capacity-building for the whole of the Federation. There were ten activists in our group. As we were starting the session, a couple of activists attached empty flip chart papers onto a wall, while the rest of us took the chairs from around a table and arranged them in a semicircle facing the flip charts on the wall. We had three questions to guide our discussion. Three or four people assumed the lead by speaking up first. Sharbil, who was sitting closest to the flip charts, took on the role of 'note-taker' and everyone mumbled their endorsement. As the discussion went along the activists who had taken the initiative and stuck paper to the wall, acted as facilitators: when one of the others wanted to make a point, we addressed them initially. In these breakout groups, however, there were no speaker lists. On the contrary, the informal facilitators actually invited those people who had not yet spoken to offer their views.

While a person spoke not everyone would face them. Some watched Shar-bil write on the flip charts, some looked at what was already on the flip charts, others reflected and did not look anywhere in particular. Our gazes, followed by heads and shoulders, even the torso or whole body, sometimes turned to look at the person speaking, at other times turned down to take notes. The tone of the discussion was collaborative, not confrontational or debate-like. The participants were facing the same direction, metaphori-cally, in the forward-looking, collaborative creation of ideas, in the same way that they faced the flip charts together. On these papers, their ideas were being made into a tangible and identifiable thing (the written points to be presented back to the plenary). It is not only that their ideas were assem-bled together on the paper that made the work collaborative. The discussion was a conversation that flowed. Every new comment followed on from the previous one, adding a new point or a different take on what had been said before.

Although face and eye contact were definitely parts of what created the sense of immediacy, they were not the only factors that generated a sense of presence among the participants. Presence in such face-to-face situations was a compound of many things, including the awareness of bodily proxim-ity, the warmth this generated, the sound of the speakers' voices, their facial expressions, gestures and the breathing of persons sitting all round, the light that surrounded the whole group (which, like the walls of the room, cre-ated a pocket in which we worked) the responsiveness of speaker and other participants (who shifted together almost as in a dance from looking at the speaker, to surveying the flip chart). These aspects were wholly integral to the sense of immediacy or presence, and most importantly, participants could *respond* with any or all of their person. Due to all these aspects, I sug-gest that this type of presence is better described as 'all-round-person' rather than 'face-to-face'. The latter term distracts attention from these other forms of responsiveness that create co-presence.[15]

The use of flip charts and the small spaces of these meetings have two important impacts: they educate participants' attention to the presence of particular activists through their voices and through their ideas written on paper; and they contribute towards changing, or reinforcing, the activists' experience of presence and immediacy, which is crucial for their daily work with ICTs (see Chapter 7). Additionally, there are two particular practices carried out by facilitators and other experienced activists in the breakout groups that educate the attention of newcomers in the understanding of inclusion and the attitude towards diversity considered characteristic, or fundamental, to 'FoEI'. The first point is, the facilitators and other experi-enced activists ensure that in *these* types of meetings everyone has contrib-uted, especially if there are first-time participants, or participants who are not confident of their language skills. Second, they encourage the partici-pants to be creative in their thinking and responding to the ideas of others—they kept asking how can apparently opposing ideas be approached so as to complement each other? This stance to discussion was easily taken up by participants, both experienced activists as well as newcomers, in all the meetings I participated in (six international meetings between June 2006 and December 2008).

The international meetings change the activists through the specific work, the joint activity, that they carry out. Through this they undergo an education of attention (Ingold 2000, 2001) which is, in effect, an apprenticeship into a particular orthopraxy. During the meetings, activists learn or reinforce particular ways of listening, of discussing, of taking decisions, and of understanding 'environmentalism'. They learn different ways of recognising the presence of particular persons. Activists get to know each other and at times they also develop close friendships. The personal relationships are important for their work too. Annamaria explained that "telephone and email are wonderful, they have really facilitated our work. But if I don't know the person writing the email or speaking on the other end of the phone it's so easy to have misunderstandings.' In these meetings, the activists get to know each other in many ways, by talking to each other and the information that they share, but also by coming to recognise the sound of their voice, the sound of their laugh, the words they use when they write and the way that they write. Like signatures, a person's specificity is imbued in these different aspects. To the extent that when the activists leave the meetings and subsequently come into contact through email or telephone, they have learned each other's particularities so much that they are present to each other in a very tangible (not virtual) sense.

Friends and Kin

Earlier in this chapter, I described how activists who develop close relationships through their meetings consider each other friends. Friendship is an important concept for the activists. FoE Malta's mission statement, collectively created in 2007, describes how FoE Malta activists strive to be 'the human voice of the earth', implying, both in their written documents and in their attitude towards their environments, a relationship not only of representation but of care. They are friends with other activists, but they are also friends with the Earth. However, at the outset of the chapter I also described how the chairwoman of FoEI welcomed the new member groups into the 'FoEI family'. Many South American FoE activists call each other 'brother' at the meetings and the leader of FoE Nigeria describes FoEI as a sisterhood and a brotherhood. It may seem odd that notions of kinship are so widespread when the title of the organisation is *Friends* of the Earth.

On the one hand, the name Friends of the Earth is not so central to the international meetings. In an interview, David Brower admits that when a group of them were "fishing around for a name", his wife came up with the idea "Friends of the Earth". They liked it and used it, but did not give very much weight to it. Moreover, since the national member groups are expected to have been set up before they join FoEI, they tend already to have an established name other than Friends of the Earth, which they keep. Finally, however, the notions of kinship invoked when activists talk about "family" are practical tools. As with the strategic thingification of 'FoEI', their use of these kinship metaphors is often an attempt to bring about the sense of closeness associated with family. In the same way that Cambridge University has 'mothers' and 'fathers', kinship is deployed to engender a more committed affiliation (Bamford and Leach 2009: 12). At the international meetings it is not clear whether the impact of these metaphors is equally powerful for the participants present. Introducing their volume on *The anthropology of friendship*, Bell and Coleman (1999) observe that it is hard to maintain a sharp distinction between friendship and kinship, noting "the power of kinship as an idiom through which to express all social relations considered to have binding qualities" (1999: 6). This power is clearly at work when FoE activists speak of "family".

However, again it is important to note that here family is not to be understood as belonging in genealogical terms, as sharing blood (Bamford and Leach 2009). Neither is it the 'deference to unity' that Plüs (1995) identifies as the source of unity in a decentralised organisation such as the Religious Society of Friends (RSF, or Quakers). Unity is still the aim of FoE activists. Feelings of solidarity, and the active solidarity entailed in supporting other groups (and not merely stating it) can be likened to the aim of unity that Plüs finds among the Quakers. The key difference is that for FoE activists, unity and solidarity do not in practice mean similarity and agreement by compromise or silence. Fighting' for a cause is valued. Indeed, the need for different degrees of violence in different situations (such as women from the 'movimento sem terra' in Brazil destroying GM crops), is discussed by activists, not rejected outright. Furthermore, that different FoE groups and FoE activists have varying, possibly conflicting ideas about environmentalism is *valued*, since it is seen as arising from different environmental situations. Solidarity and belonging are rather generated by continued co-respondence in shared experiences.

Communitas

We have seen how the meetings are bubble-like in many ways. They create certain rhythms through the joint practices of the participants and rhythms that create anticipation for future interaction (Lincoln 2011). The classic explanation for such a communal sense of belonging lies in Turner's idea of *communitas* (1979). Although there are striking similarities between Turner's (*ibid*: 238) descriptions of *communitas* and the FoEI meetings, there is a fundamental difference. There is perfect equality among Turner's initiands, which resembles the equality given to first-timers at FoEI meetings, who officially have the same status—anyone can contribute to the discussion and decision-making is mostly by collaborative consensus. However, at FoEI meetings the particular does not become general as in Turner's descriptions of the liminal phase (*ibid*: 236). The FoEI meetings do not collapse all the participants' differences as should happen in a liminal phase. Liminoids are not considered polluting (*ibid*). Finally, the distinction between the organisers and the participants is not a relationship of complete authority of instructors over neophytes (*ibid*: 237).

In addition, Turner's elaboration of the notion of *communitas* neither simply corresponds nor conflicts with practices at the international meetings. *Communitas* is:

> A direct, immediate, and total confrontation of human identities, [with which] there tends to go a model of society as a homogenous, unstructured communitas, whose boundaries are ideally coterminous with those of the human species. Communitas . . . is strikingly different from Durkheim's "solidarity," the force of which depends upon an in-group/out-group contrast. To some extent, communitas is to solidarity as Henri Bergson's "open morality" is to his "closed morality".
>
> (Turner 1966: 132)

In relation to the activists, the issue is complex. In their ideals, some of which were discussed in the life interviews and others of which are embodied in their mission statements, the activists imagine a world where the spontaneous, unstructured (read egalitarian) model of society extends not only to humankind, but to the Earth and in some cases (Anthony and Demian) the whole universe (Gatt 2009). The notion of an 'open morality' also seems to resonate with FoEI's ideology of inclusiveness and its celebration of diversity. Finally, the *communitas* among persons on this large all-inclusive scale is ideally a process that Anthony Buber calls "the life of dialogue" (quoted in Turner 1966: 143). The activists' explicit work includes deliberate efforts to refine processes of dialogue at the meetings. The emphasis on dialogue, on collaborative consensus, is one of the ways in which activists practise their ideology of inclusiveness. The understanding that *communitas* shuns property and celebrates poverty, as a means to remain outside of a social structure (*ibid*: 145) that creates inequalities, resonates with FoEI's discourse on 'grassroots'. 'Grassroots' are the poor, never the powerful. In practice, however, I have shown how in-group/out-group distinctions are sometimes drawn, even if only provisionally. And although FoEI's openness to diversity recalls the open-morality of *communitas*, that openness is limited.

One of the main reasons FoEI *exists* is to campaign against practices considered unsustainable and unjust—their current slogan reads (or is shouted) "Mobilise! Resist! Transform!" The priority of many FoE groups is to resist the activities and impositions understood to destroy the environment within that group's circle of interest. More illuminating is Turner's distinction among three types of *communitas:* 1) existential or spontaneous;

2) normative; 3) ideological (1966: 132). To illustrate the move from existential to normative he describes the works of St. Francis of Assisi, founder of the Franciscan order, and of Caitanya, the 'revivalist' of a Vaisnavite movement in Eastern India. In their visions of devotion, both strove for permanent *communitas* (for Francis the mortal life was a liminal phase in the passage to immortal life, *ibid*: 144) and in practice this meant avoiding those things that pertained to normal structured society: property (Franciscans) or love in marriage (Vaisnavites) (*ibid*: 157).

As I described in Chapter 1, the historiography of FoEI tells that when David Brower proposed FoEI in 1971, the aim was *not* to have a structure.[16] The Federation remained largely non-institutionalised until the mid-1980s, when the AGM created structured campaigns, formally elected an entire ExCom and the secretariat settled in Amsterdam. Even then, the institutionalisation was more on paper than in the obligations expected of member groups. Today, the notion that anyone could pick up the name Friends of the Earth and join the Federation has been completely replaced by a long application system.[17] With the introduction of the SVPP, nearly all FoEI practices are now structured. Where before, FoEI members acted as FoEI representatives when and where the opportunity arose, now only carefully negotiated positions become 'FoEI' positions. If Turner's concept of *communitas* can shed light on anything in relation to FoEI it is probably this, that the original vision of the founders (who actually were the hippies to whom Turner refers in his article) was more akin to an existential *communitas*, which, because of the expansion of the Federation, became normative *communitas*. Certain occasions, such as the international meetings, are instituted to provide openness and lack of structure, and thus to revive a sense of belonging that is not attributionally based, to rekindle existential *communitas*.

However, Turner's thinking belied certain structuralist undertones, and as such he takes stability and endurance to be the norm. He defines the liminal period as necessarily outside of the social structure (1979: 234–236). He defines the relations of *communitas*, being liminal, as *anti*-structural relations between 'total human identities' (1966: 132). This means that by its very definition *communitas* can only ever relate problematically to the relations between activists, due to both the highly volatile and constructed character of boundaries and groupness, and the ongoing work that has to be done to bring about the existence of 'FoEI'. Stability and structure have to be worked at, in FoEI at least.

Summing Up

The various situations that I have described so far show that activists perceive the existence of 'FoEI' as an entity of sorts, even if shifting, elusive, and only intermittently present. However, neither in the discourse nor in the daily practices of activists is it understood that the sort of thing that FoEI is has always been the same or will never change. Most types of change are neither denigrated nor celebrated. Only the endurance of some activists and their activities, no matter how small, so long as they can be called a FoE group or FoEI, count for the 'thing' FoEI to exist. This is clear in the words of Richard Sandbrook, an early FoE activist from Britain.

> The start of Friends of the Earth, and indeed of FoEI, was romantic to be sure, but it was also very hit and miss and mundane. Day by day you never knew where the money was coming from, nor who would take the slightest notice of what we did.
>
> (FoEI Archives Amsterdam)

If the etymology of the word 'thing' as a parliament[18] is given precedence, instead of its contemporary meaning as a completed 'object' (with the unintended Durkheimian connotations of the super-organic, extra-somatic and mechanical that come with talking of a human group as a 'thing'), then one way of understanding how FoEI hangs together is indeed by thinking of it as a 'thing'. It is a 'thing' constituted in dialogue, agreements and disagreements, and collaborative work with others, that changes the shape of the 'thing', making its boundaries so porous that at times they seem not to exist, and at other times to be remarkably stark. The analogy I propose, to assist in the understanding of how the 'thinginess' of FoEI may look over time, is the way that starlings storm, grouping and regrouping. Some drop out and others join in, in a cloud that constantly changes form, and yet produces a visibly distinct 'thing'. The thing does not depend on coherence or homogeneity to be distinct. In fact, were the *trajectories* of the starlings 'homogeneous', or 'the same', then their spectacular collective shape shifting would not be possible.

Interestingly, when Victoria Timmer, who wrote her PhD on the comparison between FoEI and Greenpeace, presented her work to the FoEI IS, she described FoEI as a spiral that picks up different ideas, practices and people as it moves along, and the eye of the spiral as what endures. Judith described this to me, showing me with her hands the movement the spiral takes as it moves, and expressing a deep agreement with Timmer's conclusions. However, there is a significant difference between Timmer's spiral and my description of FoEI as a thing. The spiral, as Timmer and Judith describe it, is FoEI, and the things it picks up are people's ideas and practices. The spiral is not itself the people, but their ideas. Furthermore, the spiral with a single 'eye' also imagines FoEI to have a strong centre, whereas in my fieldwork I found that different groups, such as FoE Brazil may not participate directly in FoEI, but still explicitly develop their work as part of FoEI. So in a sense there are a number of 'eyes', which in the 'starling' analogy are easy to picture as little eddies within a larger fluid mass. Nonetheless, I do not see a great difference between Timmer's spiral and a 'thing'. Indeed were the spiral to be described as a spiral storm cloud which is actually constituted by the other clouds and winds that merge with it or detach from it as it moves, I could have simply adopted the analogy.

The international meetings help bring about a very particular sort of relationship among the activists who participate. The intense sociality and atmosphere engendered at the meetings means that participants can get to know one another *personally*, and form genuine friendships which are important for their relational sense of self. Thus, activists become 'the person I know' to each other, rather than, say, 'the national representative for Malta'. It follows that what the meetings *produce* is not just a federation of nationalists (in the Durkheimian mould of organic solidarity) but a community of friends spread right across the world, within a protean FoEI 'thing'. By creating these relationships, through the experience of air travel, the images of FoEI members spread around the globe on websites and other forms of print media, FoE activists develop a sense that their activism, as well as their notion of the environment, is global.

Notes

1 See Doherty and Doyle (nd) for a map of these collaborations among all the FoEI member groups.
2 The goals, values and mission of FoEI are not the same as the mission, vision and values of FoEI that I mention in other chapters. The latter was only developed in 2004.
3 See www.foei.org, accessed on the 20th February 2008.
4 See the last sentence of this article about the protests against the extension of the development zone in Malta in 2006 www.timesofmalta.com/articles/view/20100314/local/hundreds-protest-and-call-for-better-environmental-protection, accessed 19th January 2011.
5 See for instance, the strategic essentialisms of indigenous groups referred to by Anderson and Berglund (2003), who represent themselves as a bounded group more as a recourse to their situation than because they experience an essentialised community.
6 See www.foei.org/en/resources/publications/annual-report, accessed 19th January 2011.
7 I discussed the question of whether the work of FoE groups that is not *intended* as FoEI work can still be considered FoEI with core of activists at the IS. They thought about it and agreed that since FoEI is a network of grassroots NGOs the intention or self-awareness of the multitude of FoEI activists around the world is not the point. Their group having joined FoEI, their work *is* FoEI.
8 Rapport (2002) argues how this sort of internal homogeneity is normally achieved by force of political power, while Mol and Law (1994) argue that homogenous 'regions' are created by eliding differences.
9 See Doherty and Doyle (nd) for the different definitions given to local communities by different FoEI member groups.
10 The academics were not allowed to attend the Swaziland inter-BGM, and I was only allowed on the condition that my participation was as a representative of FoE Malta—in fact, Anthony didn't attend that time. For this reason, my conclusions are also derived from that meeting, but I do not describe the Swaziland meeting ethnographically.

11 Of course, Durkheim also made the point that social facts are real (1982 [1895]), but in this thesis I am arguing against the separate domain of the social that Durkheim posited, which is the basis for his assertion about social facts.

12 Here, non-interactive websites such as FoEI's public site are part of print media. See Appadurai (1996) on the effect of ICTs on imagined globescapes.

13 See Doherty and Doyle (nd) for their map of what they categorise as the core, intermediate and peripheral FoE groups in FoEI activities. They also map the collaborations among different FoEI member groups.

14 The use of the word 'bubble' was also the topic of a running joke in Abuja. The strategic maps, which took three days of different forms of discussion, had various ideas, all individually contained in 'bubbles'. By the end of the fourth day of the BGM, after arduous discussions about the specific wording and implication of each 'bubble', all the strategic maps of different levels had been endorsed. Most activists joked that they had had enough 'bubbles' for a lifetime.

15 Lee and Ingold (2006) also show that the notion of face-to-face and eye-to-eye, along the lines that Simmel expands on, is confrontational rather than companionable. As with the walkers Lee and Ingold worked with, FoE activists develop their sociality with each other by going along together in a common environment, at least when they are at the international meetings.

16 Recall Brower's words "We made it a point not to be clearly organised or directed by some old tired formula from the top. Find good people with the right ideas and let them move ahead their way." (FoEI Archives Amsterdam, 1995)

17 When new groups apply, they spend two years as associate members with invitations to discussions but no voting rights, and in the meantime they are vetted by other FoE groups in their 'region' and members of the Secretariat, who visit their operations.

18 See Latour (2003: 54) and Ingold (2007), who both define 'things' not as readymade objects but as relations. Coincidentally, both refer to the etymology of the word 'thing' as lying in public process and discussion.

7 Communication Technologies and Presence

Being in Touch in FoEI

In 1982, the editor of FoEI's newsletter, *Link*, explained how a number of FoE groups listed as members had not replied to any mail correspondence for several months, and he was considering them defunct. He then called on any activists going on holiday to the countries where these defunct groups were from to visit the offices, either to confirm that they had ceased to operate or to re-establish communication. Almost thirty years later, this still occurs. A member group stops responding to mail correspondence or attending meetings, and the responsible members of the Secretariat try to find out whether this lack of response means that their group has folded. However, the means available to the Secretariat and many member groups to get in touch with each other have grown to include not only face-to-face meetings and mail, but also cheaper telephone calls (local, regional, national and international), fixed-line conference call facilities, email, online document repositories, text messages through mobile telephony, instant messaging and voice-over Internet protocol (IP) services, such as Skype. Especially in the last ten years, these new means of communication have radically changed the way FoE activists work together, how they work with their fellow activists within the same national member group, how they work on international campaigns or working groups and how they interact with the International Secretariat.

In other social sciences (Horton 2003; Leeder 2007; Washbourne 1999, 2001; Doherty and Doyle 2009; Pickerill 2001), and in social theory (Castells 1997, Wilson and Peterson 2002; Harvey 2000; Giddens 2007), environmental movements are among the social movements identified with social change due to their use of new communications and information technologies. These writers have mapped the use of mobile phones and the Internet in relation to the growth of environmentalist practices and social groupings. Their writings refer to transnational environmentalist actions and groupings, such as the demonstrations in London, Seattle and Prague, which were largely organised and promoted using new ICTs including mobile phones, as well as to environmental actions and groupings that work within smaller areas (Horton 2003). Likewise, the national FoE member groups with which I carried out fieldwork (FoE Malta and FoE Brazil) have wholly incorporated new communications technologies into their daily work. In the international work of the FoEI federation, ICTs are also thoroughly incorporated into daily practices. The ExCom and the Secretariat constantly work to include those member groups who do not have access to regular electricity supply, or reliable Internet, or reliable fixed-line connections, by organising regular face-to-face meetings.

In anthropology, as yet, there is relatively little research specifically about environmental NGOs and even less about their use of ICTs. However, ICTs have been gaining increasing attention in anthropology since Escobar's review in 1992. Much of this literature addresses two main issues: person-hood (Turkle 1984; Correll 1995; Reed 2005; Boellstorf 2008); and the massive social change towards a 'Network Society' induced by such ICTs, forecasted in scholarship by Manuel Castells (Green, Harvey and Knox2005; Garsten and Wulff 2003 and the contributors to their volume, Miller and Slater 2000). The work on both of these themes focuses on the social prac-tices *surrounding* these technologies, and they diagnose either that ICTs are subsumed into already existing social forms or that they create specific changes in certain social forms of the users. In this work, besides Marilyn Strathern's (2000) and Michael Jackson's (2002) more theoretical writings, notions such as 'virtual' and 'virtual co-presence' remain vague. We are not offered possible explanations of the processes by which such technologies are successfully incorporated into people's lives. Since for FoE activists, ICTs are so central to their work and lives, to their national FoE group and FoEI, this chapter is dedicated to exploring how activists engage with and through ICTs. FoE activists experience 'presence' in their use of different communi-cations technologies, and by focusing on this aspect I propose to address the vagueness that surrounds notions of 'virtuality'.[1]

In the literature I refer to, presence is also called co-presence or imme-diacy. In the first section of this chapter, I explore FoE activists' engagement with different communications technologies. Such interaction through ICTs is often distinguished in the literature from face-to-face interaction by the adjective 'indirect' (for instance Escobar 1999; Calhoun 1991, 1995).[2] In fact, when many FoE activists engage with each other through communica-tions technologies it is experienced so much as direct presence that it would be empirically incorrect to refer to it as 'indirect' or 'at a distance'. There-fore, in the second section, I list the practices that make a presence through ICTs, and analyse how presence appears in different situations. I go on to describe how the presence co-generated with ICTs makes it possible not only to maintain but also to develop personal relationships. In the fourth section, I describe the processes during face-to-face meetings whereby FoE activists learn to perceive the sense of co-presence through ICTs. In the fifth section, having touched on some situations in which ICTs do not so easily co-generate immediacy, I discuss by whom and for whom such technologies are developed.[3] I explore the mystifications these discourses create. Presence co-generated with ICTs is actual and not virtual presence, and the discourse of virtuality has political implications. The discussion of ICT-mediated presence leads me to question the conventional understanding of co-presence, referred to as 'face-to-face'. So to conclude, in the sixth section I analyse the notion of face-to-face to explore what creates presence in *these* situations.

Uses of Communication Technologies

Naturally, the work practices of activists differ from group to group in FoEI. These differences extend to the use of communications technologies. Some practices involving ICTs seem to be more widespread, while others are restricted to particular groups and situations. Common to all was that the different means of communication they employ were seamlessly interwoven with other ways of working and interacting in their daily tasks (Calhoun 1995 and Agre 1999:4 also note this). The technologies most used are telephone, email and Skype. During meetings where activists are in metric proximity will refer to emails, phone conversations and online documents. In a Skype chat, the activists may refer to a particular meeting held in metric proximity, or a document being circulated by email. The work and interaction carried out by means of different modes of communication are all enrolled as a matter of course in activists' daily lives. They specify what they are referring to—'we said this, or that, during that Skype call'—for clarity, not to highlight a barrier or discontinuity between the different modes of communication.

When a new technology is available (such as when Mariangela, FoE Brazil, downloaded Google Earth for the first time), or when reflecting on past ways of working in FoEI, the ICTs used today are generally praised by the activists. This 'techno joy' (Garsten and Wulff 2003) can be seen in the email I received from Elisa. She wrote proudly telling me that she is now on Skype, even if she had originally *not* joined because she considered overwhelming enough the ways people already had of getting in touch with her (email lists, individual emails, texts, phone calls and meetings in metric proximity). Many activists have mentioned the heavy workload that they see connected to the increased access to information through the Internet, as well as people's increased access to them through more channels of communication. However, overall, different communications technologies are welcomed and become transparently part of the daily lives of the great majority of the activists with whom I did fieldwork.

Among the different ways of being in touch, FoEI activists hold a number of international meetings (see Chapter 6). These are the main meetings in metric proximity where they meet primarily as 'FoEI activists'. The FoEI website also has a section called 'Insite' that can only be accessed with a login and a password through which activists share and store documents. Finally, the FoEI International Secretariat manages the 'FoEIAll' email list, and any FoE activist can ask to be added to the list. Currently, there are around 300 email addresses on this list. However, comparatively few emails are exchanged here. Usually, these are notices about annual international meetings, calls for participation in new projects, programmes or campaigns and notices about people leaving or joining key roles or posts within the network. For instance, the executive director of FoE EWNI sent an email to this FoEIAll list to announce that he was leaving his post. On occasions such as these, a handful of people send replies to the whole list, with brief messages of farewell. Occasionally, there are also discussions on FoEIAll. When Al Gore was presented with the Nobel Prize in 2007, a press release was issued by the FoEI International Secretariat congratulating him. This led to a string of around twenty emails over one week, in which a number of people called for the press release to be withdrawn, explaining the reasons why Al Gore should not have been presented with the prize. Usually these discussions happen within different campaign working group email lists, in which only a handful of FoE members participate by comparison with the numbers on the FoEIAll list.

The campaign teams are often made up of no more than ten to twenty people, since there are only a small number of national member groups who have the capacity to work on international campaigns (see Doherty and Doyle 2009). These teams are set up to have 'regional balance' and this means that at least three of the campaigners in each team are likely to have limited access to Internet or international telephone connections. In July 2007, while I was in Amsterdam, the Forest campaign team attempted to hold a Skype conference call. Participating were two people from South America, two people from Amsterdam, one person from Cameroon, one person from Papua New Guinea and one person from Japan. I was not part of this conference call, but I was sitting in the vicinity and it was obvious from the two people in the Secretariat office who were participating in the call that it was not succeeding. Roberta repeatedly had to explain, speaking quite loudly and slowly, that there was obviously a delay for a couple of the participants, so they had to wait before they replied. In a call that lasted around an hour, two participants got disconnected at least twice each. Sometimes Roberta and sometimes Guntur said, 'Hayao's been disconnected', and later either Guntur or Roberta would welcome him back. During a subsequent staff meeting, Roberta shared the frustration she felt at how difficult it was to facilitate the meeting with those problems of delay and disconnection. Skype conference calls tend to work better with fewer participants, and to work successfully Skype requires a relatively large bandwidth. She stated that she would organise a fixed-line telephone conference for future meetings, despite its being much more expensive than Skype.

Apart from the difficulties with Skype, the international campaigns are also, politically, the most delicate working groups within the federation. In these groups, campaigners often hold a wide variety of experiences and policy positions that sometimes are difficult to reconcile. Campaigners meet, in metric proximity, at the events that they plan actions for. Therefore, for instance, the Climate Campaign team had a campaigners' meeting in Rome in September 2007, in preparation for a UN meeting on climate in Bali in December 2007. They met again for a campaign meeting in Croatia in June 2008. In between such meetings, campaigners often have individual Skype conversations, there is a campaign email list, and different campaigners keep in touch and exchange opinions and ideas by smaller email groups or one-to-one emails.

Both FoE Brazil and the International Secretariat in Amsterdam have an office and employees, as well as volunteers. The activists in these groups meet in the office a number of times a week, depending on their work programme and how many days they are paid or bound to work from the office. However, whether in the office, working from home, or from somewhere else, both offices have an internal email list for all the activists of that particular group. In practice, even if sitting beside, in front or behind someone in either of these offices, activists there receive regular emails from people in the office and people out of the office. Both the Brazilian and IS offices have a weekly staff meeting, to which people are tacitly expected to attend. Staff meetings are generally held on Mondays in the afternoon in both offices and followed very similar patterns. Furthermore, certain activists from FoE Brazil and the IS have regular and long Skype conversations with people from other organisations they are working with on joint programmes, campaigns or projects. Almost all the members of both the IS and FoE Brazil receive daily emails from a number of campaign lists, both FoEI campaigns and campaigns or discussion groups of other federations—such as the World Rainforest Movement, the International Rivers Foundation, or the Climate Action Network. However, during one staff meeting the Coordinator of FoE Brazil asked people to cross the office to speak to each other, rather than sending emails even if the person was there in the office. Otherwise, the group emails were used to allow all the activists to keep in touch and up to date, even if not in the office daily.

The FoE Malta group, on the other hand, found a usable office in 2007, and only Elisa would regularly work from there. For many years before getting this office, the FoE Malta activists worked together primarily through an email list and email discussions. Until his sudden death in June 2005, Julian, the chair of FoE Malta, would organise more or less monthly evening meetings at a bar or café. After his death, the rest of the group met more regularly to plan how to proceed without Julian.

Previously it was widely felt that FoE Malta was held together mainly by Julian's charisma and constant emails. Anthony, who was the most active member after Julian, proposed including FoE Malta in FoEE's capacity building project, Adelante. As a part of this two-year project, more meetings in metric proximity were organised and various separate email lists were created. Most activists on lists send replies to the entire list, however there are a number of activists, also depending on the situation, who prefer to answer the sender of the email, rather than replying to the list.

There are different types of emails. Dave Horton (2003) classified the different types of emails sent by environmental activists in Lancaster into informational emails, outreach emails and reinforcement emails. Though valid for FoE emails too, there are also a number of different kinds of emails that do not fit comfortably into Horton's categories. Here is a more comprehensive list:

- Informational emails: sending articles, emails from other networks, notices of meetings, events, documents being published. As Horton described for Lancaster environmentalist emails, the senders of some, but not all of these emails, would simply write 'FYI' (for your information) in their email before the attached information email.
- Planning emails: Normally someone sends out a notice of an upcoming event. These notices are sometimes considered informational emails, depending on whether the activist who receives them can or wants to participate in the event. Often these event notices are followed up by emails where the activists discuss who will or can participate in the event, or who can or wants to follow the issues addressed by that event.
- Discussion emails: with or without attachments—discussion of press releases, feedback on policy documents, including documents for local government, regional government, national government, or international bodies such as the UN, CAN, FoEI positions etc. . . . The documents attached will have comments inserted by different activists as part of the discussion.
- Introduction and farewell emails: introducing new members, possible new volunteers, campaigners, farewells from long-standing activists, congratulation emails.
- Call for help or support emails.
- Personal emails: these often mix work with personal and the tone is entirely different.
- Translation emails: FoEI employs activists from REDES (FoE Uruguay) to translate emails into Spanish. As yet, only important FoEIAll documents are translated into French as well. Possibly there are programmes in which I didn't participate (such as the forest campaign, where more African groups participate) that have more French translation emails.

A volunteer with FoE Malta or FoE Brazil receives around ten emails a day that they are expected to read. Activists involved in one international campaign and listed on FoEIAll, receive around twenty emails a day. An activist, who is involved in a number of different groupings—FoE Malta, FoEE ExCom and a number of FoEE campaigns, FoE MEDNET, FoEIAll and a FoEI working group—receives more than forty emails daily, but they do not respond to all of them. Anthony has proudly told other FoE Malta activists that he has organised and stored all the emails of FoE Malta, FoEIAll and a number of other lists since 1999. In all he has archived over 10,000 FoE emails. Many of the activists I engaged with during fieldwork were at least following two or more of the international campaign lists, as well as other email lists from other environmentalist networks.

Except for Elisa, currently FoE Malta's only employee, I have not come across many activists using their mobile phones for work conversations. Texts are sometimes used during actions, normally to check where people are or what stage of work they have reached, to get other people's telephone numbers, and among other things to remind people of work to be done. Although the people contacted by mobile phone may include many FoE activists, the large majority of mobile phone calls, texts and Multimedia Messaging Service (MMS) are used for personal and non-work reasons. In fact, although FoE activists' ideal distinction between work and non-work is mostly blurred in practice, it sometimes becomes clearer through their subtle unwillingness to spend their money on mobile calls for work.

Although I had little experience of using instant messaging online with FoE activists, I did observe in both the FoEI IS and in the FoE Brazil office that while on Skype or otherwise working on the computer, the activists who spend the most time dedicated to FoE and also who have friendships with people who live in other cities or countries, also had a couple of chat windows going. Mariangela, Andrea in Brazil, Bettine, Imee, Roberta and Guntur in Amsterdam had these chat windows going throughout the day. They would not be chatting all the time. The chat window would be minimised most of the time, its icon soundlessly signalling in its own particular way when a message had been received. Mariangela or Guntur would then spend a few moments exchanging messages before minimizing the window again and returning to other windows that were open on their computer.

The topics of these chats varied widely. Guntur and Emanuela, sitting in the same office, once spent a couple of weeks using IM to organise a camping holiday in Bulgaria together with a couple of other FoE activists not in the office. Before the camping holiday, there was a FoE event in Bulgaria, so the holiday was planned around that event and the chats to organise the trip mixed of discussion about the event with details for the camping holiday.

Anthony has regular Skype conversations with Roger from FoE EWNI, the twin FoE organisation assisting FoE Malta in its capacity-building project. I had a number of Skype conversations with Anthony to prepare for international FoE meetings while I was in Aberdeen, Brazil and Amsterdam. Anthony told me that the first thing he does when he gets up in the morning, sometimes before and sometimes after breakfast or getting dressed, is check his emails. Anthony's computer is set up on a desk in his bedroom in his parent's house. Most activists, who also work from home on occasions, also have computers set up close to their sleeping quarters. Judith's desk and computer are in a little study within the bedroom area and floor of the four-storey house she shares with her husband, her three children and two other tenants. Mariangela's desk is in front of her bed in an alcove surrounded with Indian-style printed cloths, books, documents, incense sticks and various objects she has been given or has collected herself. Not only is

Figure 7.1 A long-standing activist employed in the Amsterdam IS office: computer; telephone; headset.

the ability to communicate through ICTs important in the activists' daily lives, the place where this communication happens is itself made important by the location of the computer and the things it is surrounded with. ICTs are clearly central to most of the working practices I observed during fieldwork (see Figure 7.1). It is also clear that activists' engagements with each other are not impersonal 'working practices'; they are also about friendship and often become very personal indeed.

Exploring Different Types of Presence

As others have already noted, communications technologies are invisible except when they break down (Jackson 2002; Latour 2005). For most FoE activists these technologies are so much a part of their daily lives that they are 'transparent'; the activists do not *need* to pay attention to the technology in order to use them. A growing number of anthropologists, but also scholars in media studies, social psychology and sociologists, have focused on the place of such technologies in people's daily lives (Miller and Slater 2000; Horst and Miller 2006; Morse 1998; Turkle 1984; Urry 2002, 2003; Horton 2003; Washbourne 1999). Their work generally counters deterministic notions that these new communications technologies would usher in a revolution of social structure with global proportions. They show, rather, that the technologies offer the possibility of modifying the lives of people who use them, but are not independent of the social practices and contexts within which they are used. The theorist Craig Calhoun (1991, 1998) proposed that in order to understand the changes occurring in relation to these new technologies, researchers should start with communities[4] and explore how communications technologies are integrated into their ways of life. The works cited earlier do exactly that. However, they continue to oppose the kind of inter-subjective interaction and communication they call face-to-face to technologically mediated communication. Something is either face-to-face or it is mediated, at some sort of metric distance by technology.

Since interaction by means of communications technologies is experienced as a *distinct type* of presence, presence itself requires further understanding.[5] Indeed Urry (2003, 2002), referring to Knorr-Cettina and Bruegger's work with stock market traders, speaks about the possibility that 'co-presence' might gain new dimensions through new communications technologies.

That activists are present to each other by means of communications technology is evident in at least three different ways:

- Participation by telephone. This is considered 'official presence' for some meetings that require participation for the fulfilment of duties in certain official posts.
- Email discussions, where a press release, a policy or a document is sent to a group of emails and then comments, feedback, amendments and discussions are carried on by people responding to the whole group. This involvement is considered to be participation to the same degree as is round-table discussion. This makes it possible to include people from different FoE member groups and people who have different office schedules.
- Many activists develop intimate friendships or working relationships based on a mixture of meetings in metric proximity, email and telephone interaction.

By exploring in detail a few of these types of presence, through reconstructions based on fieldnotes, headnotes and memories, some key aspects of how activists are present to each other will become apparent. These examples are not intended to illustrate a general prototype of presence. Rather, each example shows the different affordances of presence that are generated in practice in particular situations.

Email Lists

For many years, the FoE Malta activists worked together primarily through an email group. To date, since the death of Julian and the completion of the capacity-building project, email has been complemented by more and more regular meetings in metric proximity. Currently, FoE Malta has two main email groups—one is called *harruba* (harruba@foe.malta . . .) and the other *dielja* (dielja@foe.malta . . .). *Harruba* is the Maltese name of the carob tree, indigenous to Malta, and as such has become considered part of 'national Maltese' identity (Gatt 2001). At first this email group was formed by people who were active in FoE Malta before the start of the capacity-building project, but now it includes only elected board members and Elisa.

The *dielja* group, which is the Maltese word for a grape vine, is for the wider group of activists and volunteers. When, during the capacity-building project, more meetings in metric proximity were being organised, the meetings for all activists were called '*dielja* meetings'. The email group and the meetings in metric proximity were inseparable. There is a distinction between the types of presence, but those who participated in one were automatically involved in the other.

The *dielja* and *harruba* email discussions have their own rhythms, different from metric proximity or telephone conversations and interaction. Real-time replies are not necessary. For certain issues—such as feedback on documents, or uncontroversial press releases—activists are given a date, usually around three or four days' time, by which to send comments. At other times, activists just happen to be checking their email when someone sends an email to the group and their reply is immediate. This happens frequently because most of the activists either work or study in a situation where they use both computer and Internet regularly. Anthony and I have frequently commented, in almost real-time email conversation, how strange the conversation feels, how immediate. Here, email becomes more like telephone and instant messaging, in the sense of mutual responsiveness.

Otherwise, the fact that email does not require immediate response is one of the aspects that makes it so successful in FoE Malta and other volunteer-based groups. In previous fieldwork carried out with Nature Trust Malta (NT), Cedric, a general medical practitioner and medical researcher and a core member of the organisation, explained to me how email changed his involvement in NT. He was then often out all day, either seeing patients at his clinic or out on house visits. In the past, his fellow NT members would contact him by telephone and pager. He felt that these forms of communication disrupted his work and he could not give the intrusions, as he considered them, much attention. With email, however, no matter what time he gets home, he can read the emails without rushing, he contributes to NT discussions and brings himself up to date on events and opinions (Gatt 2001). It is the same in FoE Malta for all the activists who are in full-time employment, or have families to care for, even for those who participate only occasionally in FoE Malta. Email allows them to catch up with discussions and information, up to many months later, and it is possible to refer back to details of discussions, although the sense of presence is stronger when people respond within the day.

Joint Documents

Email group discussions are also central to the work of FoEI's international campaigns. The international campaigns, set up in 1982 with more or less the current structure of a lead group (FoEI Link, Amsterdam Archives), were initially based primarily on annual coordination meetings. The campaign groups communicated through limited postal correspondence or meeting at events to keep in touch between Annual General Meetings. With the proliferation of email, these lead groups (now called international programmes and campaigns) came to depend less on meetings in metric proximity and interact a lot more by email. However, as FoEI expanded and included more groups from 'Southern' countries, with less access to Internet and telephone, and with the SVPP's new prerogative to have 'regional balance', the reliance on electricity and electronic communications technology has had to be reduced and regular meetings in metric proximity are organised more often. Nonetheless, email group discussions are still common. The International Climate Campaign mailing group sends an average of fifteen emails a day. Many are translations either from English to Spanish, or vice versa.

The FoEI campaign team group emails include all the different types of emails I listed in the previous section, except the more personal emails. However, both proposed changes to documents and policy discussion by email are longer in the international emails than in either FoE Malta or FoE Brazil emails. With documents attached, the discussion process in the international campaigns is also longer than in FoE Malta.

FoEI is a member of the International Climate Action Network (CAN), which is regularly invited to contribute documents to the UN's intergovernmental climate meetings. The representative of FoEI on CAN first circulates the draft documents prepared by a coordinating body in CAN. In the case of the document circulated in March 2008, a FoE EWNI activist who attended a CAN meeting volunteered and took it on himself to circulate the CAN documents and call to the other members of the climate email group and to compile all the comments into one document. (see Figure 7.2) Once the email group members received this compiled document,

Figure 7.2 The CAN document with comments from various activists on the International Climate Campaign, worked on by the email group.

Note: The comments are blurred for confidentiality.

more email discussion began about the various comments added to the document. In this case, after many long emails over not more than 24 hours, in a discussion that went on between five FoEI activists including the former chairperson of FoEI, it was decided that the writing team could submit the document with comments because in the light of UN Kyoto negotiations it was central to FoEI's Climate Campaign. However, this decision was on condition that the document was not the final position of FoEI on the Kyoto protocol. The actual 'FoEI position' was discussed in the next meeting held in Croatia in June 2008.

The documents shared through email discussions such as these create their own records and histories. Instead of needing someone to keep minutes, as in meetings held in metric proximity, every time an activist leaves a comment in the document their amendments show up in a different colour, automatically including the name of the person to whom the computer is registered. The same goes for email discussions. To double-check the wording of a proposal, or to check who said what, all that needs to be done in an email discussion is to scroll down the page. As a result, the detailed wording of press releases or policy documents, or of organisational documents such as the statute, are left for email discussion, leaving the principles and the main thrust of the arguments to be carried out in meetings in metric proximity.

Staff Meetings

At the International Secretariat in Amsterdam, Griet, the executive director of the office, works from Washington. To fulfil her duties as director, Griet works on documents at home and participates in email discussions, she travels to Amsterdam for executive committee meetings and for other annual office meetings (such as the office retreat or annual staff reviews), she discusses projects with individual staff and with groups by telephone or telephone conference and finally she participates in the weekly staff meetings by means of a telephone conference device. This device transmits voice in analogue, which makes it more audible and, most importantly, reduces delay. The reduction of delay in telephone conversations and conferences is crucial because one of the experiences that most successfully creates presence is the perception that there is no unusual gap between the moment an interlocutor speaks and the moment he or she is heard.

During staff meetings, the staff in the office at the time would sit around a large oval table. Most times, there were at least nine people around the table. The telephone conference device, a flat circular microphone/speaker, would be placed in the middle of the table. When the meeting was due to start one of the staff would press a button to call Griet. A dial tone would be heard and Griet saying 'hello'. The first part of the meeting was generally a round of reports from the staff and volunteers about their work in the previous week and their plans for the next week's work. During this round, the device was pushed around the table to be closer to the speaker to help Griet hear well. She often could not hear us that well. In the Amsterdam office, the activists addressed each other and the device (Griet) when speaking. The listeners however shifted between looking at the speaker, their notebooks while they took notes or reviewed them, or at an unfocused spot on the table. In these telephone conference meetings, the voice is the most important aspect of a person's participation and facing each other is of lesser in importance, even among those in the same room.

Imagine that, contrary to what happens during these meetings, the participants *did* face each other rather than gazing at some unfocused spot. Immediately, the situation would become conspirational. Present participants would be communicating, through unspoken facial gestures, a sub-text from which the (metrically) absent Griet would be excluded. So, because of Griet's presence only through the telephone, participants are almost compelled not to look at one another in the eye, for fear of appearing to plot behind her back. Although only one person is actually metrically distant, in order to maintain the inclusivity of the meeting, participants have to act *as if* they all were. Different communication technologies not only create actual presence, they also change the nature of face-to-face presence. Therefore, relationships can also be developed by interacting through ICTs.

Personal Relationships

Roger Rouse (2002) was one of the first anthropologists to write about the use of the telephone as an integral part of how Mexican migrant communities in the United States maintained their thick social relations despite being dispersed in different geographical locations. A number of sociologists have also researched the role of email in maintaining social relations. In his study of environmentalists Lancaster, Horton (2003) argues that while email works very well in maintaining social relations, in Lancaster at least, it was not very successful in attracting new activists.[6] In FoE, email and telephone have been very successful in maintaining relations, in ways similar to those described by Rouse (2002) and Horton (2003). In addition, I came across a number of instances in which new activists were recruited through their websites. However, the contact that some activists develop through a mixture of email, Skype and meetings in metric proximity entail more than simply maintaining social relations. These activists developed relationships that were both professional and friendly by getting to know each other more and more over the years; their friendships changed and developed through the various ways of being in touch.

Nadia and Mariangela first met in 2004 as members of an email group of the REDE Brasil (Brazil Network).[7] In 2004, Mariangela was elected joint coordinator of the network. That year Nadia had just been employed as the secretary of the network. Mariangela was based in the office of NAT in Porto Allegre in the southeast of Brazil, while Nadia was based in the office of another NGO in Brasilia. Mariangela and Nadia corresponded by email every day, both working out the responsibilities of their new posts and preparing for various meetings. Among other responsibilities, Mariangela and Nadia had to jointly organise the annual network meetings, as well as ensure that the other members of the network carried out the tasks they took on. By 2006, when Mariangela was still joint coordinator of Rede Brasil and Nadia still the secretary, they had become very close friends and colleagues. They regularly spoke on Skype during the day and evening, not only discussing political developments, issues related to Rede Brasil and NAT, but also such non-work issues as their relationships with their partners, their health, their families and their hopes for the future. They went on camping holidays together in the summer with their partners. When Mariangela visited Brasilia she stayed with Nadia. Nadia also stayed with Mariangela whenever she visited Porto Alegre. Their relationship developed over the years. When they first met in 2004, they were not already going on holiday together or staying at each other's homes. The many dimensions of their relationship grew through their regular contact that included meetings in metric proximity as well as regular encounters by means of ICTs.

Many activists in FoE groups are hosted by other activists whom they do not at first know, while travelling, who then at times become close friends. In fact, although not all activists develop relationships that are both work relationships and friendships, this type of relationship is very common among FoE activists. Imee and Roberta, Judith and Roberta, Roger and Anthony, Anthony and Karl, Anthony and Anastasia, Felicity and Lili, Kathia and Nely, Marisa and Emilia, Imee and Guntur, Emanuela and Bettine are only a few of the close friendships that I have got to know better during my fieldwork. I present this list of names only to convey a sense of the number of close, and sometimes intersecting, friendships and personal relationships that arise from these varied modes of communication. All these relationships developed through a mixture of relations in metric proximity (including international meetings, holidays, visiting, travelling or going to places together, at times living or staying at each other's homes), regular email, text messaging and telephone or Skype calls.

Unpacking Presence

For FoE activists, hearing the voice of a person and conversing with them over the phone, whether they already know each other from encounters in metric proximity or not, results in a trustable context for communication, an immediate presence. With email, on the other hand, it is not hearing the responses of other FoE activists that creates presence but their email addresses and very personal turns of phrase. Both a name signed at the end of an email as well as the email address from which it is sent are essential aspects of how presence is constituted with emails. Both come to have the same personal identification with the writer that Ingold (2007a: 94) has suggested in the case of a signature.

Unlike a signature, which becomes a part of a person's identity by being so deeply engrained in motor memory (*ibid*), an email address becomes a part of a person's identity because—with the growing system of personal logins and passwords used in online personal identification—it is assumed that only that particular person has access to their own private email. Often the name at the end of the email only confirms this, because from my experience and my observations with various FoE activists, the sender of an email is remarked by their name or address listed in an email inbox. Even before the actual email is opened, the receiver already knows they are opening an email from a particular person. Many activists explained to me that they visually filter their email inbox for the names of people either from whom they are expecting word, or who stand out, rather than opening up all emails consecutively. Subsequently, when reading the email itself, the words printed on the screen are not anonymous, but the words of the person who sent it: Stephanie calling for comments; John sending a clarification and explaining that he had been ill all day; and so on. As I have mentioned in the previous section, activists regularly refer to emails while talking on the phone or during meetings in metrix proximity. Often, if not always, when activists refer to an email from someone they talk about what so and so *said*, not what so and so wrote. Taken together with the value of email discussion being equal to round table discussion, this simple reference to a person's voice shows how closely activists associate the written email with the voice of the person who is sending it.

Not only FoE activists read the words of fellow activists as though hearing their voices. In the medieval period, perception of the written word was modelled on people's experience of hearing those words spoken (Ingold 2007a: 13). According to St. Augustine, 'when a word is written it makes a sign to the eyes whereby that which pertains to the ears enters the mind' (cited in *ibid*: 14). When monks read, they would pronounce the words, murmuring as they read and in so doing conjure up the 'voices of the pages' (*voces paginarum*) (Carruthers 1998: 170). Carruthers writes about how within the medieval philosophy of being, the bodily performance of writing and the intellectual comprehension of those written words when reading were perceived to be as closely related as eating and digestion. As such, the *voces paginarum* could not be dismissed as figments of the imagination. Rather, we should pay attention to the ways that the monks' disciplined and practiced experience of hearing the words read or sung, and of reading or singing them themselves, can leave a trace in both aural and muscular consciousness (Ingold 2007: 15). 'To read, then, is not just to listen but to remember' (*ibid*). Writing in the medieval period through to the Renaissance was a way to connect with the past. It provided 'pathways along which the voices of the past could be retrieved and brought back into the immediacy of present experience, allowing readers to engage directly in dialogue with them' (*ibid*).

When FoE activists read each other's emails, the *voces paginarum* do not belong to deceased saints but to living, responding fellows, and these voices are remembered from previous meetings in metric proximity and telephone conversations.[8] In addition, email is particular, because for the activists it does not follow strict rules or formal writing styles. This allows people to write emails in ways very similar to their styles of speech. This is still a matter of surprise to some activists who comment on how, when reading each other's emails, they can hear the other person talking through their distinctive discursive style. In Chapter 6, I argued that during the international meetings, participants come to associate other people's ways of talking with their writing. At the meetings, participants can feel each other's proximity through a variety of modes (by seeing them, feeling their warmth and their smell, or the pressure of their proximity through the air and by hearing them). With this knowledge of each other, and of the way emails are written, co-presence is as vivid as it is over the telephone. In the same way that people new to the telephone learned to recognise people only by their voice (Marvin 1988), now email users are also learning to recognise people by their style of writing. Though there is still a difference, since voice is recognised on the phone by prosody rather than choice of words alone, whereas with email we recognise a person by the choice of words and not by the scribal equivalent of prosody, which would be the person's handwriting.

The immediate presence of the other activists is confirmed not only by the deeply ingrained memory of their voices but also by their prompt email replies. Different from email and beyond simply memory, the real-time of telephone and IM (and the possible real-time of email) draw the activists' attention to the immediacy of presence afforded by these communication technologies; equal in *value* to, but distinct from, presence in metric proximity.

When Presence Is Not Afforded

FoE activists experience actual presence when using communications technologies in the different ways I described previously. This presence is co-produced by the FoE activists, their experiences, the technology (the particular machines they use and the infrastructure that supports them) and other aspects of the environment together with and in which they interact. Presence thus emerges from the co-engagement of the different constituents of that environment. The distinct constituents that come together to create actual presence become apparent when something breaks down (Jackson 1983; Scarry 1994).

At the Secretariat in Amsterdam, Emanuela was responsible for helping with the logistics of organising international meetings. Consequently, Emanuela was one of the main activists preparing the inter-BGM to be held in Swaziland in September 2007, helping the national FoE member group with the reams of logistics involved in organising a FoEI meeting. It became immediately obvious that Yonge Nawe (FoE Swaziland) had virtually no access to email or little familiarity with working by email. For this reason, Emanuela would telephone FoE Swaziland at pre-organised times, since they did not have someone working in the office full time. These telephone calls were not easy. Emanuela would often spend more than an hour on the phone. The line was more often than not very unclear and she would have to raise her voice for her interlocutor to be able to hear her. Many misunderstandings also arose from these conversations, specifically regarding the logistics of visas and accommodation and the planned duration of the pre-conference.

In the end, one week before the meeting was planned to start, Imee, who was then in South Africa, was convinced by the rest of the office to travel to Swaziland to help with the organisation of the meeting. It had not been possible to offer sufficient support by telephone and email, as had been done the previous year with the meeting in Nigeria. Members of the Secretariat and activists from FoE Nigeria at the Swaziland meeting nevertheless fondly remembered how even when organising the meeting in Nigeria, the Secretariat and those members of ERA organising the BGM had to plan the times to send emails because there was only a short time window when there would be electricity.

In this example, a combination of factors prevented the communications technologies from establishing sufficient presence for the various problems that arose to be solved. These included FoE Swaziland activists' lack of access to regular, fast Internet, lack of access to stable telephone lines and lack of experience in coordinating a project through communication technologies. Similar 'blocks' were involved in the case of the Forest Campaign conference Skype call that I described in the first section. The particularities of how Skype works, where each interlocutor requires considerable bandwidth for the quality of speech that Skype achieves, means that even with broadband Internet, Skype copes better with small conference calls. Another example occurred during the BGM in Nigeria in 2006. Due to the lack of a stable electricity supply, and broadband Internet, and the poor mobile phone reception (caused by the relief of the landscape and fewer mobile phone relay points), the activists felt that the meeting was being held in a bubble, cut off from the people and the relations they are accustomed to maintaining (Gatt 2009: 115).

There is also a difference between people who regularly use communications technologies in their daily work and those who are less familiar with email, telephone and Skype, even if it is available. As I have shown, activists need to be educated into the potential presence their use of communications technology affords. The affordance of presence conferred by the use of communications technology is not self-evident, nor does it replace other types of presence. It is important to note that affordances cannot be equated with 'objective characteristics'. Rather an affordance, and subsequently an effectivity (the effectivity of presence afforded by the activists' use of communications technology), emerges from the activists' relations in practice (Ingold 2009, 1992). Therefore, if certain aspects of the relationship are changed, the communication technology may no longer afford co-presence. For instance, when the telephone was introduced to Amish communities in the United States at the turn of the last century, there was debate among their leaders about the effect of this new communications technology on their way of life. The result, after a faction broke away from the main Amish denomination, was that the leaders banned telephones from Amish homes on the grounds that the maintenance of accountability and solidarity depended on contact in metric proximity (Zimmerman Umble 1992).

This is a reminder that communications technologies are built by particular people with particular experiences and interests. The telephone and the Internet were both developed in situations where voice and writing already held the potential for co-presence. Evidence for this lies in the widespread notion that sound, and therefore voice, personifies (Ingold 2000a: 272); the notion of personhood inscribed in a person's writing, their handwriting or signature (Ingold 2007a; Carruthers 1998); as well as familiarity with the notion of witnessing through writing established by Boyle's written experimentation practices (Shapin and Schaffer 1985). The telephone and the Internet were developed in a context where, unlike in Amish understandings, communicative practices could serve purposes other than building community solidarity.

The fact that the presence afforded by the use of communications technologies is not self-evident, or necessary, does not reduce the *actuality* of presence that FoE activists experience. The instances I described earlier show two things. First, when co-presence through the use of communications technologies is achieved, a whole host of co-engagements co-produce this effect. Second, FoE activists experience their interaction through communications technologies as co-presence despite the fact that it is not a *necessary* product of these technologies. Thus, the affordance of presence is learned. Through their daily use of communications technologies, the FoE activists learn the ways in which it can afford distinct forms of co-presence. Finally, participation in metric proximity is recognised as different from an information and communication technologies (ICT) presence. As I have mentioned, the presence constituted with communications technology is not confused with, and does not replace presence in metric proximity. In fact FoE groups, including the FoEI ExCom, invest considerable amounts of time, money and energy holding meetings in metric proximity. This is done not only because many groups do not have the opportunity to use communications technologies on a regular basis but also because there are particular aspects of presence in metric proximity that cannot be achieved by email or telephone.[9]

Virtual Presence?

In media and cultural studies, Margaret Morse argues that communications technologies produce fictions of presence.[10] She argues that these fictions in television and computer-supported 'cyberculture' are *not* 'less authentic than "real" discursive exchange between human beings' (1998: 10). Rather, basing her argument on Berger and Luckmann (1971), she claims that 'socially constructed reality is already fictional and that virtuality is an aspect of that fictionality that has come to be more and more supported and maintained by machines' (Morse 1998: 11). This is not the place to critique the premises of Morse's vein of social constructivism; others have done so convincingly elsewhere (Ingold 1992, 1993, Latour 2003; Berglund 1998). However, her focus on 'discursive exchange' as the determinant of social presence is telling. The sociologist John Urry (2003) also bases his description of face-to-face co-presence on processes of conversation. By focusing narrowly on discursive practices as prototypes of social interaction and consequently of co-presence, these approaches miss fundamental aspects of sociality, which also shed light on why communication technologies have been so efficient in developing and maintaining social ties, not least in Friends of the Earth.

I have described how through their interactive practices, FoE activists—together with particular communications technologies—co-produce different types of presence. These types of presence are not fictions created by the suspension of disbelief that Morse (1998: 18) compares to the fictionality of film and television. To show why, it is necessary to clarify the rhetoric implicit in the notion of 'virtuality'. When not referred to as *indirect* communication or interaction, people's interactions through communication technologies are said to be *mediated*—for instance by computers (Wilson and Peterson 2002; Escobar 1994; Corell 1995; Gray and Discroll 1992) or telephone (Horst and Miller 2006; Rouse 2002). Thus, 'mediation' is used to distinguish between online and offline interaction (Castells 1997), or between virtual and physical presence (Escobar 1994), or between response and embodied presence (Knorr-Cetina and Brueger 2000). In every case, the assumption is that in so-called 'offline', 'physical' and 'embodied' interactions, presence is *not* mediated. Knorr-Cettina and Brueger (2000) carried out research with stock market traders in different cities. By analysing the computer-mediated practices of the traders, it became apparent that the other traders were 'present' to them. They recognise that the computer-mediated real-time responses to each other's actions were central to this presence. To capture this kind of presence, they coined the notion of 'response presence', as opposed to the 'embodied presence' of persons in metrically proximate contact. What they fail to recognise, however, is that 'response presence' depends on the capacity of transduction in electronic communications technology to facilitate these real-time responses. Transduction is the conversion of variations of energy, produced for instance by a voice, into electrical (with electric cable) or optical (with fibre-optic cable) variations, and back again. Thanks to transduction, FoE activists participating in the telephone conferences I described previously, and stockbrokers negotiating deals, are just as present to one another—that is, as fully embodied persons—on the phone as when they are listening to presentations at a meeting in metric proximity. For Knorr-Cetina and Brueger to oppose '*response* presence' to '*embodied* presence' has the paradoxical effect of *disembedding* persons in metric proximity from the surrounding medium. For even when people are metrically close to each other, perception and interaction are still mediated—not in the sense that sensory perceptions are assimilated to predetermined cultural models, as Geertz (1973), Berger and Luckmann (1971), or Strathern (1991, 2000) would have it, but in the sense that our actions and perceptions are produced in tandem with the medium we inhabit. We hear, see and feel only because we inhabit a medium, namely air, which transmits radiant light, whose vibrations conduct sound, and whose currents brush the surface of the skin. Different media—such as water or earth—with different transductive properties, afford correspondingly different qualities of perceptual experience. However, this applies with equal force to the media that are enrolled in the operation of communications technologies. What air is to encounters in metric proximity, electric or fibre-optic cable is to

ITCs. Sights and sounds shared by persons in metric proximity, through the mediation of air, are no more *physical* than pictures and voices shared by persons who are metrically distant, by way of wire or fibre-optic cable. Even in machine-mediated communication, the people involved in the interaction are fully embodied and are physically present to each other.

During their almost daily Skype conversations, Julia in Amsterdam and Griet in New York are *themselves* as embodied as if their conversation were happening in a meeting in metric proximity. The co-presence Julia and Griet perceive in such cases *does* depend on real-time responses, as the failed Forest campaign Skype conferences and the difficulties in organising the meeting in Swaziland showed. Therefore, although Knorr-Cettina and Bruegger's notion of 'embodied presence' is problematic, their notion of 'response presence' is as useful in describing stock market traders' practices in different cities as it is in describing telephone, Skype, and video or webcam conversations in FoE. The notion of 'response presence' also helps us to avoid having to attribute such presence to the existence of a 'virtual space' that people enter, leaving their bodies behind them. The uncritiqued notion of virtuality remains wide-spread even in the anthropological literature on communications technology, and I argue below that this happens when the environmental, material and non-human aspects of such engagements are not taken into account.

Whose 'Virtual' Reality?

The vagueness surrounding 'virtual reality' was already present with the spread of the telephone at the turn of the last century. Then, electricians used metaphors of travel to describe how the telephone works so as to make themselves understandable to a general public, or so they thought. They spread a metaphor that depicts 'words . . . carried by electricity' (Marvin 1988: 90). The resemblance to more recent depictions of the electronic is clear: computers carry us into new worlds, virtual worlds. In reality, electricity does not *transport* people's voices (or text, or information); rather, the process, as I describe above, is one of transduction.

The term 'virtual' remains vague in contemporary research on the Internet. Even in their efforts to show how the use of communications technologies is embedded in social and political contexts, many anthropologists (Miller and Slater 2000; Garsten and Wulff 2003) do not explore how interactions are actually achieved through ICTs. For example, drawing on their research in Trinidad, Miller and Slater (2000: 4) see the Internet neither as a placeless space (cyberspace) nor as a means of disconnection from particular places. Nonetheless, they retain the notion of 'virtuality' to distinguish online from offline interaction. Garsten and Wulff (2003: 2) adhere to Miller and Slater's position that even practices carried out only online are connected to daily life. Yet to show how such online interaction is not detached from offline sociality they suggest the notion of 'social virtuality', without questioning the latter term.

Much of this anthropological research (Miller & Slater 2000; Hasslestrom 2003; Green 2003; Rapport 2003; Wulff 2003) focuses on people's practices. Similarly to my presentation of the very real type of co-presence experienced by FoE activists, Wulff argues that '[v]irtuality is as real and social as any life form' (Wulff 2003: 200). It is curious that Wulff and others should retain the notion of 'virtuality' to describe the relations formed through ICTs without unpacking it. Green, Harvey and Knox (2005) begin to unpack the rhetoric of the 'virtual world' promised by the EU's promotion of ICTs. They posit that 'physical space' and 'digital space' are probably part of the same thing, and that if 'print capitalism became entangled in generating imagined communities of nation-states, as Anderson argued, then it is likely that the new technologies will also be used in political place-making projects' (*ibid*: 806). However, their work focuses on those who have been cut out of those 'digital places', and they do not explore the constitution of such places in detail.

In an address to the conference 'Virtual Society? Get Real!' (2000), Marilyn Strathern outlines how from the seventeenth century the term 'virtual' came to refer to the effects of characteristics (virtues) on their own, without the need for the 'actual' virtue or characteristic. Virtual means that effects have efficacy in and of themselves; they are auto-enabling. She goes on to argue that it is only because the Internet is *perceived* as a type of auto-enabling system that it can also be perceived as decontextualised, that is, as 'virtual'. Here we can see what is ultimately part of the 'standard view of technology', according to which technology is understood to enfold within itself inherent and autonomous forces of progress (Pfaffenberger 1992).

Issues in the philosophy of technology are not directly tackled in most of the anthropological work I have reviewed on communications technology, though as I note earlier, many argue against the technological determinism implicit in, for instance, Castells's (1997) writings. Escobar (1994: 211–212) outlines the development from the conventional understanding of technology as autonomous from society through to the constructivist view of technology (in STS) that emphasises the contingent and flexible character of technological change. He attributes this shift in the understanding of technology to events in the 1960s that increased awareness of the negative effects of industrial development, and the nuclear industry in particular. In his review, Escobar (*ibid*) traces how anthropology came to 'cyberculture' mainly through the work of science and technology studies (STS) scholars. This social constructivist entry point possibly explains why so many anthropologists interested in 'cyberculture' and communications technology failed to attend to the material and ecological, which underpin the human uses of these technologies. In parallel with STS, Ingold (2000a) also argues against technological determinism. However by basing his argument on an analysis of the relations between the human body and tools and machines, he lays the groundwork for a very different approach to the anthropological analysis of technology, and consequently of communications technology, than the research of Miller and others. By failing to pay attention to the material relations that enable Trinidadians, for instance, to maintain their Trini sociality across oceans, Miller and Slater (2000) do not explain why the Internet was so successful, while radio for instance was not. Furthermore, they do not expose the international players such as American Telegraph & Telephone (AT&T) or Alcatel, which lay and maintain a number of the world's undersea optic cables. If these cables are missing or broken, the Internet is experienced very differently.[11]

According to Woolgar (2002), the notion of the 'virtual' in computer-mediated communication is part of the political economy of such technologies, where increased power comes from the *invisibility* of the human relations involved with them. Gray and Driscoll (1992) and Stone (1991) draw our attention to the very real human interests implicated in their development. They trace how the development of 'virtual reality' technology, the Internet, imaging technologies and other communications technologies were funded by and for the U.S Military (Gray and Driscoll 1992, 42; Stone 1991). Despite their origin, such ICTs no longer serve only military ends; in fact, as we have seen with the cases surrounding Wikileaks, these technologies are often used to subvert those interests. However, others have shown how the rhetoric of 'virtuality' is a key notion in political and economic projects (Green, Harvey and Knox 2005; Miller 2000; Green 2003; Escobar 1994). Thus, in addition to an empirical motivation to explore and explain how presence is constituted, relations described as 'virtual' deserve fine-grained analysis due to issues of social control in the far-reaching networks of capital and power within which they are imbricate.

Presence in Metric Proximity

The notion of 'being there' is an established part of anthropological fieldwork due to the underlying assumption that sociality is ultimately constituted in 'face-to-face' situations (Gupta and Ferguson 1997; Amit 2000). The focus on 'face-to-face' interaction is based on the importance given to vision and eye contact. In the works on social interaction of George Simmel ([1908], cited in Lee and Ingold 2006: 79) and Goffman ([1963] cited in Urry 2003) mutual visibility and eye contact are the primary means by which individuals are entwined. According to Urry (2003: 164), it is specifically in 'face-to-face' interaction that 'conversations are produced, topics can come and go, misunderstandings can be quickly corrected; commitment and sincerity can be directly assessed.' However, others have argued that the co-presence so central to sociality (Urry 2003, 2002; Hastrup and Hervik 1994), is not dependent on, nor simply a matter of, face-to-face or eye-to-eye contact (Bird-David 1994; Lee and Ingold 2006).

In their chapter on walking and fieldwork practices, Lee and Ingold (2006: 79) point out that the sociality of walkers does not depend on their interactions being face-to-face. Rather the sociality of two people walking is an inherent part of their shared rhythms, their shared vistas and exchanges in conversation, all of it involving their whole bodies in the process (*ibid*). The sociality of the FoE activists as they work together in the plenary room, or in the workshops rooms that I described previously, or indeed as they hold Skype conferences or jointly work on documents by email, are also involved in a sociality that does not depend on strictly eye-to-eye nor face-to-face interaction and communication. The success of interaction— successful turn taking, the generation of trust, correction of misunderstandings and so on (which Urry 2003, following Simmel and Goffman, ascribes to eye contact)—are achieved by means of the whole body. Posture is perceived not only through vision, but also through air pressure, or bodily pressure or the warmth of people sitting close together. The perception of sincerity can be gleaned from the tone of voice. A person can be identified through the association of their person and their ideas with their written words. Therefore, in consonance with Lee and Ingold's argument, FoE activists perceive each other's presence and thereby constitute their sociality with their whole bodies and not solely by eye-to-eye or face-to-face mutual visibility. Furthermore, direct eye-to-eye contact ignores the importance of peripheral vision in the way people are aware of their surroundings.[12]

In fact, it is the very availability of the different senses in sociality—not only vision and direct eye contact—that make telephone and ICTs so successful, although even with the telephone people learned to distinguish between separate senses. Marvin (1988: 91) recounts that when the telephone was introduced to the mainstream in the United States in the last decades of the nineteenth century, people were mostly concerned about being seen. People had to learn to dissociate being seen from being heard. The argument is similar for email and instant messaging. As I argued in a previous section, FoE activists, and in the past others such as medieval readers, perceive the presence of other persons through their written words, their personal email addresses or chat name.

Therefore, what is unique about metric proximity, in the medium of air, is that people have the possibility to respond to each other with all the means available to their person. At one of the international meetings, such as the meeting I described in Chapter 6, the FoE activists could choose to look at a person speaking, get up and sit closer to someone, or go for walks together in the cornfields behind the hotel during the lunchbreaks. Conversely, during a Skype conference call, as a participant, I could only hear the voices of the other participants. For this reason I simply describe such presence as 'in metric proximity' to replace the notion of 'face-to-face', because so far we have not developed technology that allows us to perceive and employ as many possible channels of engagement and perception as air, in metric proximity, allows us. Presence in metric proximity denotes what is currently, a unique type of presence where people can respond to each other and to their environment with many more aspects of their person than ICTs currently allow. So far, developments such as video conferencing and mobile video conferencing allow people to go for walks together and see each others' surroundings as though they were walking side by side. In mobile video conferencing people can see and hear not only the other person, but *along with* the other person. What mobile video conferencing cannot yet do is transduct smell, taste, peripheral vision, air pressure or touch.

Nurit Bird-David (1994: 598) also argues that the Nayaka of South India build their 'immediate relationships' on a proximity that is not 'face-to-face'. She bases her critique of 'face-to-face' relationships partly on Schutz and Luckmann, who hold that the category of face-to-face is too vague. Although Schutz also recognises the role of the eyes and of glances in interaction (cited in Urry 2003: 164), Schutz and Luckmann (cited in Bird-David 1994: 598) prefer to describe the basis of social interaction in terms of a continuum, ranging from 'pure immediacy' to 'pure anonymity'. Sociality, or 'we-relationships', are built upon a mutual experience of immediacy. Such immediacy, however, is eroded by distance in 'time, space or objectification' (*ibid*).

As I have argued in the previous sections, in many instances, but especially in the experiences of FoE activists, through interactive practices with communication technology, metric distance does not necessarily erode the immediacy of presence. Beyond that, Bird-David's analysis resonates with the notion that immediacy is not, or not always, determined by visual contact or eye contact. However, in positing immediacy and anonymity on a spectrum, Schutz and Luckmann rule out the possibility of different types of immediacy. Their spectrum, indeed any spectrum, depends on fixed characteristics for the definition of its poles or extremes. Depending on whether there are more or less of these characteristics you are moved towards one extreme or the other of the spectrum. In Schutz and Luckmann's spectrum, if there is *more* temporal or metric distance between people, the situation is considered to be *closer* to the anonymous end of the spectrum. Yet consider the presence, the immediacy of the *voces paginarum* of medieval readers, or the FoE activists in email discussions, who are distanced from their interlocutors in time and place. In these examples, *more* distance does not entail more anonymity. A spectrum is one-dimensional, in defiance of the multidimensionality of empirical experience. Understanding presence as having many different dimensions, such as presence in metric proximity, response presence and the types of presence of writing such as in email, or the *voces paginarum*, comes closer to what we observe empirically.

To sum up, I suggest that the types of presence achieved using communication technologies are based on more than memory. For FoE activists the real-time of telephone, instant messaging and email are distinct types of actual, not virtual, presence. After all, tools change our perception of the world by offering new effectivities (Ingold 1992), new possibilities for action, perception, thought, imagination and so on. Furthermore, as the failures of communication I have described show, these types of presence are not self-evident or necessary consequences of using such technologies. Rather the activists' attention is directed towards affordances of presence in the process of using these technologies. In Chapter 6, I described how at the international FoEI meetings, the atmosphere, rhythms and ways people worked together in groups and in rooms also work to educate activists' attention to associating people, their selfhood and their presence with their voices, their ideas, their ways of speaking and writing and by extension their written words. Different materials and different technologies have different affordances when they participate in the practices such as I have described with FoE activists. In this case, as part of the engagement with communications technologies, the affordance here is of presence, and actual presence is the effect.

I have argued in this chapter that communicative interaction is always mediated, in the sense that it is always launched forth and enabled in a medium. In this sense, mediated does not mean that something comes *in between* the persons communicating. Rather it means that interaction is possible in the first place only because persons are already immersed in a medium and importantly that the properties of that medium are bound to condition the quality of interaction. Fibre-optic cables or electric cables comprise a particular medium in just the same way as do air or water. They, too, have their own peculiar transductive properties that condition the quality of interaction. The criticism of the literature on ICTs, then, is that both in talking about 'face-to-face' *and* in talking about computer-mediated communication, anthropologists have tended to ignore the 'medium'.

Notes

1 Here I am not referring to electronic technologies such as flight simulators and other virtual reality technology.

2 It is necessary to clarify that when Calhoun and others cited earlier talk about indirect communication, this refers to whether the communication is mediated by technologies or relationships. It does not refer to indirect communication in the sense, for instance, of 'misdirection' of meaning, explored among other things by the contributors to Joy Hendry's (2001) edited volume.

3 It is necessary to note that such technologies are being developed and disseminated quickly. Wilson and Peterson (2002: 451) note how rapidly interfaces such as multi-user domains (MUD), object-oriented (MOOs) and usenet became obsolete. Since the end of my fieldwork, video-conferencing and image search engines have been developed further, although they have not yet become an essential part of FoE practices to the same extent, and with the same speed, that Skype calls and emails were adopted.

4 Here, this is Calhoun's term. I have discussed community and belonging for FoE activists in the last chapter.

5 Interestingly, Calhoun (1991) defines at least four types of relationships, from primary relationships, which are face-to-face, to quarternary, which are relationships between an individual and a large-scale system in which the individual is kept from recognising the existence of such a relationship—such as phone tapping and surveillance and more recently the detailed monitoring of people's web surfing for marketing purposes. In this scale, face-to-face and present relationships may not be the most important, or important at all in people's lives. In this chapter, although I focus on 'presence', I do not disagree with Calhoun's

position. As with Anderson's (2006) 'Imagined communities', absence may at times shape what happens in people's lives more concretely than present relationships. However, in this chapter I explore how ICTs create a presence that is significant in the FoE activists' lives and in the constitution of FoEI.

6 Timmer (2007: 42) also doubts that new technologies can foster commitment and mobilise as effectively as face-to-face and local ties (Warkentin 2001).

7 Rede Brasil is a network of Brazilian environmental and development NGOs that acts as watchdog over the workings of various regional and international investment banks such as the World Bank, the International Monetary Fund and the South American Development Bank.

8 For a more detailed discussion of the voices of the pages in relation to collaboration, see Gatt 2017.

9 I discuss the distinct processes and practices of group rhythms and other aspects of presence in metric proximity in Chapter 6.

10 In these communications technologies, Morse includes television, radio, cinema and computer-mediated communication, but not the telephone. On the other hand Carolyn Marvin (1988: 6), in her history of modern communications technologies, includes the telephone because although it is not usually considered a mass medium it was the first communication technology to enter people's homes and to unsettle customary ways of dividing the private person and family from the public.

11 See for instance, the numerous online documentaries of underseas cable faults and the subsequent loss of Internet for many countries. Malta was among the fourteen countries that were cut off from the Internet in 2008 and this prevented the activists participation in planning a meeting: http://patdollard.com/2008/12/underwater-cable-communication-cut-off-again-in-the-mediterranean/ and http://gcaptain.com/maritime/blog/undersea-cable-repairs-are-underway/ and in Asia after the quakes of 2006 http://news.bbc.co.uk/2/hi/asia-pacific/6211451.stm)

12 Interestingly, the shared vista Lee and Ingold (2006) describe in walking practices may also be present, albeit in a different way, through ICTs. Take, for instance, Griet during the weekly staff meetings by telephone conference. She often described what was going on outside her window. We could often also hear what she described, such as when we heard chain saws pruning the trees outside her window. In this way, through a mixture of hearing and imagination, Griet's surroundings were also present to us in Amsterdam. With mobile phones, although rarely used as part of the activists' work, people can be walking through their surroundings and describing them to each other. Imee and Guntur used to take photos with their camera phones and send the images as MMS to show them who was at maybe a World Rainforest Movement meeting, or where they had arrived when late for evening beers.

8 The Effectiveness of Structure

Vectors at Work in a Transnational Federation

Distance, of varying sorts, and a lack of familiarity create the effectiveness of structures. In institutions, societies, networks—any coming together of relationships—the multiplicity and particularity of these relationships do not appear from a distance. Distance seems to create an opaque surface that gives the impression that the continuous shifting, changing and flowing relationships within it are the actions of a single, coherent and—most importantly—powerful being: a leviathan (Hobbes), a reified fetish (Taussig 1992), a thing (Latour 2005). As I argued in Chapter 2, the metaphor of vectors within fields of forces adds to the notion of the world as an unfolding mesh of relations. What they add is an explanation of how some relationships have more effect than others, and a way to recognise the different qualities of relationships through the notion of fields of force. What Bourdieu calls symbolic power is of a different order from other forms of power such as access to resources (Eriksen 1998). Both of these are different from the sort of existential power and agency to which Latour and Ingold draw our attention. The strength of the analogy of vectors and fields of forces is that it makes it possible to imagine these different qualities of power simultaneously.

Furthermore, vectors make it possible *both* to take into account the intentional and non-intentional practices of persons and non-humans, their agency, *and* to identify situations when that agency is less effective. Understanding structural power to be the result of a lack of direction of attention (which may result in different forms of distance), makes it possible to take seriously the experience of FoE activists when they come up against what they talk of, and behave towards, as impersonal entities such as governments and systems. It allows us to take into account instances where things such as 'a government', 'an office' and 'a network' have effects *without* having to cancel out the agency or the intentionality of the activists (as Giddens 1984 and Bourdieu 1990 imply).

In this chapter, I aim to explore this question ethnographically. Furthermore, I use the metaphor of vectors in fields of forces as a tool to illuminate the way actions, choices, events and things relate to each other in FoE activists' lives. This, I argue, allows a better view of how the people and environments have both active effect and are affected, sometimes more sometimes less, always in different ways, by 'things' that seem independently powerful and alive.

A question immediately arises regarding different notions of effectiveness. I am using the notion of effectiveness in an ethnographic manner—that is, with regard to how the people I worked with judged success or failure. However, these people differed among themselves in their backgrounds and ways of life, so I am offering these notions of effectiveness, structure, endurance, opacity, and unprotected backs as my own hybrids of what I have learned from fieldwork as well as from the literature and from discussions with colleagues. These notions, in effect, arise from my experience. They are ideas that I hope will be picked up, in the knowledge that as notions travel, like stories from persons to persons, they will acquire new or different relevance. In what follows I describe how some FoE activists have encountered 'things', 'offices', 'governments', 'networks'. I prod these stories with my metaphor of vectors in fields of forces and draw them together in an attempt to give an account that includes both the micro- and the macrophysics of power, as the FoE activists themselves indeed experience power, powerlessness, effectiveness and ineffectiveness.

Structures Becoming Less Opaque

Izabela is a young, sharp and fashionably dressed employee at the FoEE office, working as a campaigner in Brussels. Her learning experience over the first few years of her work shows a clear, if maybe extreme way in which the notion of direction of attention can help to explain the effectiveness of 'structure', a thing whose impenetrability and therefore its power dissolves (to a degree) when attention is directed towards 'it'. The various EU bodies in Brussels appeared at first to Izabela to be impassable. The appearance of solidity had a real effect on Izabela's behaviour towards people supposedly empowered by the 'EU body' and the 'offices' they held; she was petrified by them almost into inactivity. The FoEE office[1] is oriented towards influencing decisions in these EU fora. Working in the FoEE office, Izabela found herself in a milieu where attention was specifically directed to understanding how the 'EU' decision-making structures work in order to influence these decisions. As she learned her way around and built relationships with different people, her attention was attuned to what was going on in Brussels in a way that allowed her to penetrate that previously incomprehensible 'whole' and thus to join in those goings-on.

At the EU AGM in Hungary in May 2007, Izabela presented a proposal to the plenary for the FoEE to initiate a campaign on nanotechnology. This was the first time she had proposed a new programme to the General Meeting. Until her nanotechnology research and proposal, Izabela worked on other people's programmes, campaigns and actions—but all of them, like the rest of the FoEE office, gravitated more or less closely towards EU affairs in Brussels. Izabela had particularly worked on REACH and Waste debates in the EU parliament's Environment commission and the various standing and working committees.

Her nanotech proposal was not supported by the plenary. Waste, GMOs, climate change, corporate accountability and tracking EU funds were more important at the time. So, in the evening over wine and beers, Rainer, from FoE Germany, who until 2006 had been the Director of the FoEE office for almost ten years, went out of his way to encourage Izabela. He described to a group of us standing together at the bar:

> You should have seen Izabela when she started [working at the FoEE office]! She was terrified of picking up the phone and speaking to almost anyone. Now look at her, she knows exactly whose door to knock on in the parliament, she knows which lobbyists to talk to [to get her documents included in a committee meeting], and you should see her walking straight up to important executives!

Izabela later explained to me that when she joined the FoEE office she felt 'tiny' besides the huge Brussels buildings, housing the EU Commission, the Director Generals' offices, the MEP's (Member of the European Parliament) offices, the different head offices of many multinational corporations, and the people who worked there, endowed with a power that made her feel insignificant. However, as she slowly made more and more phone calls, encouraged by her more senior colleagues and her need to keep her job, she got to know specific people and their ways. She learned how to avoid certain official lobbyists; she knew which lobbyists were paid by Shell for example. She learned which of the porters would allow her into the area where the lobbyists have their offices, where she could drop off FoEE documents and get to know the different lobbyists better. She got to know a handful of lobbyists who would then inform her which committee meetings, among the numerous permanent and ad hoc committees, were relevant to her work and when they would be held. Gradually, the huge EU machine in Brussels no longer appeared to be a single solid being, whose power infuses all those who work within it. Izabela says she now recognises where the different negotiations of power happen and can affect them in small ways. She no longer feels impotent. She proudly explained, that same night, how one of her documents was discussed at an EU committee meeting on the Waste Directive, and a small sentence from that document was included in the working policy paper. This success, for Izabela, was translated into a success for FoEE.

Izabela and the FoEE direct their attention specifically to working their way through the jungle of offices, committees, different levels of decision-making and the people that animate them, that together constitute the EU legislative bodies. In terms of fields of forces, this is a possible way of *combining* direction of attention (as shown in Figure 8.1). As I have argued in Chapter 2, in the metaphor of vectors that I am using, the notion of direction and

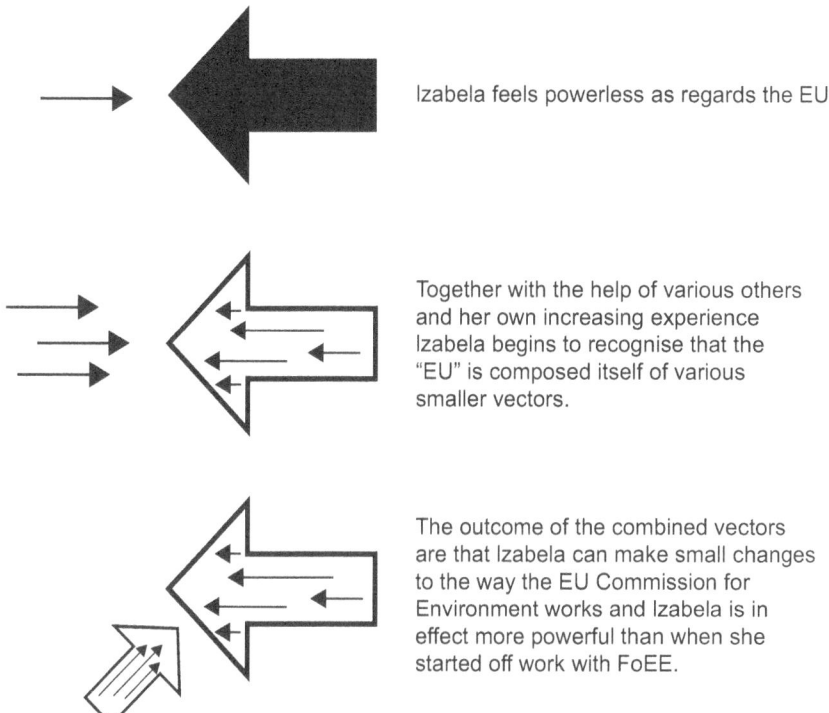

Izabela feels powerless as regards the EU.

Together with the help of various others and her own increasing experience Izabela begins to recognise that the "EU" is composed itself of various smaller vectors.

The outcome of the combined vectors are that Izabela can make small changes to the way the EU Commission for Environment works and Izabela is in effect more powerful than when she started off work with FoEE.

Figure 8.1 Structure becomes less opaque.

magnitude of force is not to be understood as a summing up of 'individuals'. By combined direction of attention, I do not mean that the more individuals combine the more force they will necessarily have to create change. Other factors need to be taken into account, such as, for instance, that different positions within already existing fields of power may have more clout than others, as well as the difference between functional power and symbolic power (Eriksen 1998: 41–45). Different notions of personhood and effectiveness, power or agency are also involved here. In the metaphor of fields of forces, vectors can grow. For Izabela, learning the workings of the EU machine and building relations with people there, the 'structure' of the machine no longer appeared so opaque as when she began her work with FoEE.

From Less Opaque to More Opaque and Strategies to Reverse That

Whereas the work of the FoEE office and Izabela's learning experience contribute towards the notion of vectors working *with* each other, Mariangela's story opens up the possibilities of vectors working *towards* or *past* each other.

Mariangela has worked with FoE Brazil since 2002. She was first employed for two days a week as a press and communications officer, having graduated in journalism that year. However, she has dedicated more and more time and work to FoE Brazil and today she is fully employed, having found her own funds. As she works on the daily administration of the organisation, participating in recruiting new employees, caring for the actual office space, she works on research and policy documents from a journalistic as well as a 'communication' angle. She is also a coordinator of the nationwide *Rede Brasil* (Brazil Network), within which she works on monitoring and campaigning on the Brazilian federal state's policies and plans. Her most exacting task is being both treasurer on the Conselho Diretor and manager of the day-to-day as well as annual finances of FoE Brazil. Taking care of the finances includes keeping detailed accounts, and chasing up the other employees and volunteers for receipts and the accounts of their actions, formulating work contracts, paying various membership fees including those to FoEI, paying salaries, setting up and following up any bank transfers and finally chasing up the IS for the Membership Support Funds (MSF) they receive from FoEI. The latter must come through in time for Mariangela to pay employees' salaries, including her own, regularly at the end of every month. This rarely happens. For at least two weeks of every month, Mariangela's main worry was: will the funds come through on time? Based on her experience of paying salaries Mariangela worked out that when there are delays, they are normally caused by someone in the IS office in Amsterdam not processing the payment forms promptly. The international transfers take at least a week. So if the 'finance person', as she called them, does not process the forms at least a week before the end of the month, they then will definitely arrive late in Brazil and Mariangela will have to deal with all the employees in the FoE Brazil office asking when they will get their pay.

From more than a week before the funds are due in the FoE Brazil bank account in Brazil, Mariangela begins by sending an email to the Secretariat 'finance person' to remind them to process the payments. When Mariangela finally receives the email confirming that the payment has been made, she then goes to the bank every day to see if the payment has been received. Sometimes the banks cause delays. Once the money is received, she pays the salaries, checking every day at the bank that the salaries have actually been sent out, then keeping detailed accounts of the dates and how much money was received and paid.

For the first few years that Mariangela was the 'finance person' at FoE Brazil, she used to send emails to 'Stine'. The two had never met, but over the years had built up 'some sort of relationship', Mariangela used to say. The relationship was based on a mutual understanding of ways of writing emails. Mariangela feels that she struggles with writing emails in English. But after three years of correspondence she had come to be confident that 'Stine', this person to whom she only had contact through an email address, Stine@foei . . . , understood her particular way of writing (see Chapter 7 on personal presence through communications technology). In 2006, she stopped receiving emails from 'Stine' and the emails were instead coming from 'Diana' (running after membership fees before the BGM in Nigeria, September 2006). When Mariangela wrote emails to 'Diana' however, she did not reply in the same way that 'Stine' had done. 'Diana' rarely responded directly to Mariangela's emails, most often sending other emails some time later with no reference to Mariangela's previous emails. Some payments were almost a month late.

In exasperation, Mariangela called me several times to check the English of her emails to 'Diana'. She worried that since Diana did not know her as well as Stine did, she might think her emails were rude, or that they were unclear. Mariangela began feeling blocked in relation to what she called '*foi*' (to sound like 'soy')—the acronym for Friends of the Earth International (FoEI) pronounced as a word. Her actions no longer seemed to have effect. She would send emails, but would not receive a reply. The emails she did receive from Diana were either of a general nature, sent to all the FoE groups, or—if specifically to Mariangela—they contained no more than one or two short lines. In fact, it turned out that Mariangela began referring to 'Diana' as the 'finance person', whereas before she simply spoke of 'Stine'.

Mariangela began exploring different strategies for 'getting through' to Diana. As I mentioned, she asked me to check her English. In addition, she wrote more emails than she used to. She wrote to Stine, too, asking for help. She considered telephoning, but then didn't; she felt even less confident in spoken English.

In terms of vectors, Mariangela experienced a growing opacity of the Secretariat based on how well, or poorly, she felt she could communicate with particular persons there and how effective her actions were. Mariangela had little familiarity with the IS; she only had relations with Stine, and then only by email (albeit over a period of around three years). When that one channel between Mariangela and Stine closed, because 'Diana' became the 'finance person', Mariangela's efforts to make something happen, to have the payments made on time, did not work. Since Mariangela was only familiar with Stine, she had no alternative points of entry into the IS. That '*foi* office in Amsterdam' became a solid, immovable thing. Even though Mariangela directed some of her attention towards the *foi* office, the magnitude of that particular vector of her attention was not sufficient to build familiarity and thus knowledge of where and how her efforts would have the effects she desired.

Her final strategy (while I was there in Brazil) was to send out an emissary to build new relations with 'Diana'. She asked me to get to know 'Diana' and find out what was happening. Why was Stine no longer her email contact? Why were the payments delayed so often, and so on? By the time I got to the Secretariat for fieldwork, Mariangela had left NAT Brazil to start her PhD in Politics and International Relations at Buenos Aires.

Mariangela's efforts are exemplary of a vector that is directed towards a 'thing', but does not have enough magnitude to influence the vectors of

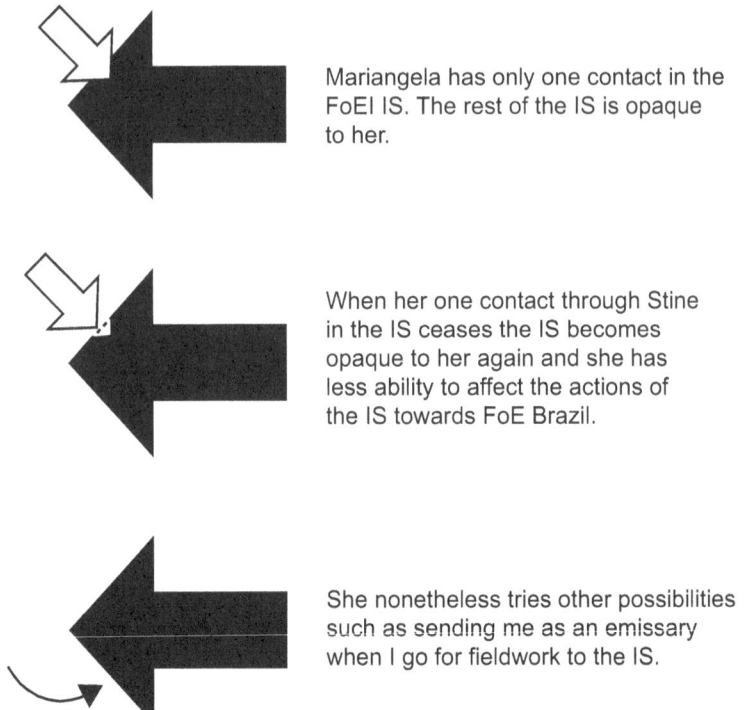

Mariangela has only one contact in the FoEI IS. The rest of the IS is opaque to her.

When her one contact through Stine in the IS ceases the IS becomes opaque to her again and she has less ability to affect the actions of the IS towards FoE Brazil.

She nonetheless tries other possibilities such as sending me as an emissary when I go for fieldwork to the IS.

Figure 8.2 The FoEI IS becomes opaque and immoveable when a personal contact ceases.

that 'thing' (see Figure 8.2). She tries to penetrate this 'thing' (the IS) to have the MSF money paid on time. Her point of access to the IS is through one person. So when that person changes the whole 'thing' becomes solid again—a goliath. Mariangela directs her attention to making the IS visible and understandable. She attempts various strategies: writing in English, getting her English checked, writing less often, more often, writing to her old contact and finally sending out an emissary, myself in this case, extending her relations through someone with whom she already has relations. As this example shows, when they do not have familiarity, activists in fact have less effectiveness in the direction they hope for—either working past each other or against each other. It does not mean they do not try.

Vectors Move Past One Another, Structures Remain Opaque

Six months later, when I went to do fieldwork with the IS in Amsterdam, I met Diana. It turns out that Diana is also unfamiliar with the different FoEI national groups; they appear opaque and solid to her. At the beginning of my stay, Diana had as yet met no activists from national groups besides the members of the FoEI ExCom, who under-emphasise their national member-group belonging at the Secretariat (I discuss this in more detail next), and myself. She had to manage the financial relations, by email, between the IS and seventy-seven member groups. She also had to ensure that all her work was consistent with the work of the others in the Secretariat. Working only two days a week, and resolutely not taking work home with her—she insisted on protecting her time to care for her adopted daughter, as well as on maintaining 'a work/life balance'—she often could not manage to do her daily work *and* answer all the emails she received in one day.

Like Mariangela towards this 'Diana@foei . . . ', Diana did not know the Mariangela on the other side of 'Mariangela@natbrasil . . . '. FoE Brazil was as opaque to Diana as the *foi* IS was opaque to Mariangela. Diana also tries different strategies to generate the effect she desires when she communicates with 'member groups' (as she refers to them) for them to pay their annual membership fees. Diana complained to me in September 2007 that around half the groups had not paid their membership fees and if they did not pay soon, this deficit would have to be presented at the BGM in November. She was concerned about this because what is presented at the GM is the official annual report and figures cannot be changed retrospectively, since the approval of such documents is by a vote of the GM.

Diana is a relative newcomer to the IS; she began working there as an employee in 2006. She does not consider herself an environmentalist because she does not participate in the actions of FoEI, but she does feel strongly about not wasting, and lives her life as much as she can along these lines. She had never been to any of the international General Meetings. However, while I was there she attended her first international programme meeting.

Although she works in the IS office, Diana also sees FoEI as an opaque thing. This is for many reasons, among them that she does not see herself as a FoEI activist and does not go into the office except for the hours of work she is paid. Therefore, during the first year of her working at the IS, she has not met the activists who from time to time drop in, nor does she join in the staff meetings, which are held on a day when she is not in the office. In addition, when she joined the office neither she nor her schedule of work was introduced by email to all those people, such as Mariangela, with whom Stine used to correspond.

Fragile 'Things'

When a Leviathan-like thing such as 'FoEI' or 'the IS' becomes known, more familiar, and less opaque it can be deployed as a 'thing' (see Chapter 6). Those FoE activists who actively promote the imagination of FoEI as a solid thing, as being separate from the agency of any of the persons who comprise it, do so because 'it', in this case 'FoEI', is perceived to be fragile.

Diana was employed to help Stine, FoEI's finance officer, because she could not cope with the amount of work involved in the currently growing FoEI Federation. Unlike Diana, Stine had worked for the FoEI IS since 2000, and before that had been director of the Secretariat of another international federation—with a similar structure to FoEI's—for two years. Stine has been to every FoEI BGM since 2000, to present the financial reports. She is a key participant in the ExCom meetings that happen every four months. Similar to Roberta (introduced in Chapter 6), Stine imagines that 'FoEI' is a constructed entity, made up from the many different ways that the national member groups work together, with the IS or simply by having become members. Stine does not expect there to be a single understanding of what it means to be a member of the network, although she does expect the membership fees to be paid sooner or later. But even in this case, if a group cannot afford to pay the membership fees, or does not contact the Secretariat for a number of years, the Secretariat will make a sustained effort to get in touch with someone from the group and if this does not work, neighbouring FoE member groups are asked to find out if the group is in need of support.

However, Stine worries that the policy documents that appear to be 'FoEI' positions (presented to United Nations working groups, or Rivers Action Network, or Oxfam Novib) may be, from her experience, tenuous and provisional agreements. She remembers all too clearly how in 2002, at the UN Earth Summit in Johannesburg, the independent position of a couple of European FoE groups clashed so strongly with the experience of the Ecuadorian FoE group that many of the South American and Central American FoE groups threatened to leave the network en masse (these were the events that led up to the SVPP; see Chapter 1). Stine conceives of FoEI as a frangible 'thing'. Being the financial officer, she is concerned that if members do not pay their fees regularly it becomes hard to justify to the funders that the members are committed to the Federation. In turn, this weakens FoEI's opportunities to raise the external funds that provide for those activities that actually do strengthen the ties between activists and member groups, that turn those relations into a more stable groupness, and so on and so forth. In Stine's experience of funding bodies and FoEI's environment of legitimation (which gives FoEI a symbolic capacity to function), presenting FoEI as a stable thing, not dependent on any particular person or persons, is one way of actually creating some sort of stability, however provisional. Stine, together with Judith and Griet, knows how to deploy the thinginess of FoEI in order to raise funds for the Federation—in her case through the financial reports for a single entity 'FoEI'.

Giddens argues that institutions are maintained by the 'unintended consequences of actions [that] form the acknowledged conditions of further action in a non-reflexive, feedback cycle (causal loops)' (2009: 14). In other words, institutions are the result of non-reflexive routine actions. However, when Stine concretises the accounts and needs of 'FoEI', as though FoEI were a solid living being which has 'needs', she does so strategically and deliberately (see Figure 8.3). She has no problem in recognising individual actions that might threaten the other relationships that are included in 'FoEI'. Consider this case: when it was proposed that FoE Europe should have a separate table at the first FoEI group 'market place' in Swaziland, she refused there and then, needing no more than a few moments' consideration. The speed with which she could gauge the consequences of particular actions and predict possible outcomes based on her experience of how other FoE groups and regional FoE groups would react showed how sophisticated was her knowledge of the workings of the 'network'. She explained, in almost exasperated tones, that 'if FoEE has a table then ATALC, and the other regions will feel left out'.

Unlike Bourdieu, Giddens grants that the division between 'practical consciousness' and 'discursive consciousness' is very porous and that things flow from one type of consciousness to the other. However, he situates the maintenance of 'institutional practices' clearly within 'practical consciousness' as *un*intended consequences of action. Yet the examples of Stine, Finance Officer at the Secretariat who works to portray FoEI as a solid entity to funding bodies, and the example of Andrea, Coordinator of FoE Brazil, to follow, show how it is possible for institutional practices to be maintained as *intended* consequences of action. Here is the place where accounts of agency (as against accounts of power) normally falter. How can some persons have the sort of agency that can shape, impose, oppress or condition the lives of others, while other persons do not? Since in anthropology, philosophy and history among other disciplines, studies of power are normally carried out from subaltern positions, such power tends to be perceived as dispersed, 'in the system', as for instance in Foucault's (1995) *Discipline and Punish* and in Lefebvre's (1991) extensive work on the power and discipline imposed by modern urban planning and architecture. In the work of both authors 'power' is omnipresent, it appears to emanate from the shapes, postures and passageways created by the architecture, as though there were no urban planners, no architects, no sponsors, no landowners. They are excluded from these accounts as ethnographically present persons.

Giddens's insistence that institutions are maintained as part of a feedback loop of unintended consequences is part of this perception of power

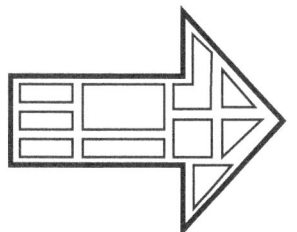

Stine collates the different forces of the FoE national member groups and deploys them in FoEI annual reports and other documents for funding bodies in a way that creates FoEI as single entity.

Figure 8.3 Deploying a fragile 'thing'.

as faceless. On the other extreme, an example of scholarship that aims to empower people is the work of Rapport. According to Rapport, 'because' causes can exist only in an account of power, which pretends to eclipse the agency of persons behind supposed structural preconditions. I argue that from both extremes, by not recognising how power over the actions of others can be an intended consequence, and by not recognising specifically when and where and why those actions (sometimes intended, sometimes not) actually manage to create 'because' reasons, the ways in which oppression and conditioning actually work (and can thereby be manipulated) remain unacknowledged. As I suggest in Chapter 2, by understanding the effect of power (intended or unintended) as the result of different vectors of direction of attention, both existential power and 'because' reasons—both of which we empirically perceive in the world—can be accounted for.

Resisting Vectors by Shifting to a Different Field of Force

The effectiveness of an opaque structure or thing can be deployed for a variety of reasons. While Stine and Roberta maintain and perpetuate the thinginess of FoEI because they perceive its fragility and coherence, Oxfam Novib puts particular pressure on the IS to make that fragility visible, at least to their inspectors.

Oxfam Novib is one of FoEI's major sources of funding. In particular, it is the institution that funds the expenses related to the SVPP (the salary of the facilitator, travel expenses for extra SVPP working group meetings, expenses for publishing, overheads incurred by the Secretariat for the increased workload). The reputation and character of FoEI as a worldwide Federation is what gives it most clout in securing funding and legitimacy with funding bodies and the political bodies and fora in relation to which the Federation is oriented (I list these fora in Chapter 4). The events following 2002, when South and Latin American groups considered leaving the Federation en masse, would have threatened this reputation—in confidence many FoE activists worried that this could spell the collapse of the FoEI. According to Imee (Members Coordinator, FoEI Secretariat), organisations such as Oxfam Novib are concerned about the lasting effects of the money they invest, such as their financial support for FoEI.

Now the accusation of the Ecuadorian group had been directed towards the International Secretariat, claiming that the Secretariat was favouring European interests and approaches to environmentalism, and was not 'International' at all. FoEI-wide policies and actions seemed to be issuing increasingly from the Secretariat, with little regard to the position of non-European member groups, especially South and Latin American groups, or so Ecuador, Colombia, Argentina, Uruguay, Costa Rica and other South American FoE groups, as well as the representatives from the Czech Republic and Denmark, maintained. In response to this crisis, the ExCom and the Secretariat leaders proposed the SVPP to the EGM in 2004 in Colombia.

Concluding statements of that EGM included the reflection that in previous GMs, too much energy had been devoted to practical actions and not enough time to discussing the internal positions of members—which explained the contradictions that led to the crisis. The SVPP was designed in order to explore and generate positions that were inclusive of all the member groups. The 'Mission, Vision and Values', the ten-year strategic plan and the implementation plans are examples of the sorts of things created through the SVPP that were supposedly acceptable to 'all the groups'. The process of the creation of these specifically 'FoEI' documents, positions, policies and plans—the very things that most often carry the reified and fetishised 'thinginess' of FoEI—had to be 'transparent' to all the groups and importantly to Oxfam Novib.

Oxfam Novib sent a social scientist to observe the SVPP process at the BGM in Nigeria. The researcher also carried out surveys after the meeting with a number of different member groups, she was in constant contact with Griet (Executive Director of the FoEI Secretariat) and seemed to fill the capacity of consultant/monitor as well as observer. The members of the ExCom and the Secretariat were so used to Oxfam Novib's attention to 'process' that interested and slightly concerned activists at the international meetings who asked me about my research questions responded with an almost standard phrase, 'Ah! You're interested in process not content [of our discussions]'. That I was interested in 'process' and not content seemed to quell whatever worries they seemed to have. Even when I protested and argued that particular processes might be linked to particular content, their worries seemed not to return: process in their eyes was entirely separable from content. The point is that the 'gaze' of Oxfam Novib on the one hand and the 'gaze' of the South American groups on the other elicited from the Secretariat in particular, because of its position in the crisis and in the structure of the Federation, *the deployment of two completely other kinds of 'thinginess' than that of a 'powerful' office.*

Julia joined the IS in June 2007 with expertise in fundraising. She was taken on to develop Griet's idea of raising funding for the Secretariat directly from individuals, simply called 'Friends'. This proposal is very controversial in FoEI because individuals are expected to support, first and foremost, their national member group. English-speaking groups are concerned that since FoEI as a federation is active mainly through English-speaking regions of the Internet they could lose financial support for their own national group to their own international federation. Julia was informed of this situation, but was not made aware of the sort of inclusive decision-making ethic that arose strongly through the SVPP between the member groups and the Secretariat and that, as a result, was also very strong in the Secretariat itself. Julia made proposals, expecting them to be accepted or rejected as wholes, and found it very difficult to understand the efforts of different IS members who wanted to contribute to her proposals and work at all stages of her idea creation process. She rejected these requests to participate as 'interference'.

For the first couple of months, this meant that she was not on speaking terms with Felicity, and had upset Lili—two members of the FoEI Secretariat in Amsterdam. Julia also felt very isolated. However, she did not leave, and the unspoken clashes with Felicity and others did not make her leave. Julia had the complete support of Griet, who found funding for Julia to become an employee within a few months of her joining the IS as a volunteer. Since Julia did not leave and slowly began to negotiate the relations with Felicity and others, she was still working with the Secretariat the following November and furthermore was presenting and facilitating the working group for a new fundraising team, comprising 'all' the fundraisers from member groups (since few member groups actually have fundraisers, this group only has around twelve participants).

On the plane to this meeting, Felicity and Lili were concerned that Julia would behave at the meeting as she did towards them in the IS, putting forward strong proposals of her own, rather than building up proposals based on the contribution of all the participating members. They tried to explain to Julia the importance of this type of facilitation, because the Secretariat is supposed to be strictly non-political (making proposals is understood in this situation as supporting specific 'interests', and therefore to be 'political'). The almost neurotic attentiveness to not proposing anything, or not seeming to propose particular positions, to ensuring that they 'have a mandate' from the member groups for any for decisions they take, as well as the extension of this approach to decision-making within their own group work, is one of the responses to the whole situation running up to and during the SVPP.

Another, similar response that I observed in many forms was to reduce the visibility of the *particularity of the persons* composing the IS. The most extreme instance was when Griet tried to erase her presence at a meeting, refusing to introduce herself in a round-table of introductions (I explore the effects of this attempt at invisibility in Chapter 9). However, other less dramatic actions have a similar intention. One of the main actions was to appoint people to the IS from very different regions of the world. Whereas before 2002 most of the employees at the IS were from 'Northern' countries (the UK, Netherlands, the United States), by 2007 the IS had employees and volunteers from at least twelve different countries around the world, including the Netherlands, Indonesia, Argentina, the Philippines, Italy, Surinam, the UK, France, Colombia, Brazil, the United States and Malta.

When describing personal identity, many members of the IS de-emphasise any single nationality. In her life history interview with me, and in many of the life history-type conversations she has had at international meetings when I was there, Lili proudly traces the roots of her multinationality family, and her experience of growing up in many different 'countries'.

> I was born in Tripoli . . . we lived there for 1½ years. . . . We lived in Guatemala for 6 months because my mum's parents were there . . . my mother's

parents—her mum is Mexican, her dad is from the USA—he lived in Mexico, first he worked in El Salvador which then got dangerous—his mother was Lithuanian. . . . I spent every summer when I was a child in Guatemala with my grandparents. . . . My father is British from Derby . . . [my parents] got married in Mexico . . . when I was 2 ½ years old we moved to Hong Kong.

I was thinking what am I, I'm not English, a lot of people tell me 'but your dad is English' but my mum brought me up. But, I also feel quite Asian, and the food, the philosophy, from the age of two family doctor was acupuncturist, the way they hold their hands, formal way of being, love to go out for meals not to stay home.

Lili speaks English and Cantonese, and is trying to pick up Spanish, which she used to speak as a child. At the two international meetings I went to with her, she would often spend the breaks talking to the representatives from South America trying to improve her Spanish.

Even Julia, who at first was oblivious to the particular pressures on the Secretariat, soon picked up the practice of effacing a 'single' nationality. At the Swaziland meeting Julia, whose parents and grandparents are all English, emphasised her life experience in Singapore and South Africa in 'getting to know each other' conversations. Roberta explained that she had been employed specifically because she embodied both the North and the South. She is Argentinian, but had lived since her childhood in the United States. She spoke both Spanish and English with equal eloquence. In the United States, she had the experience of elite political spheres, such as the United Nations offices, and she had worked for a few years as a community project manager in a predominantly black neighbourhood. In her life history interview she said enigmatically, 'I am black', although she could equally be pictured as Mediterranean or Latina with her long brown soft curls and olive skin. When in Swaziland she joined the performance prepared by the ATALC region, she was admonished by Griet and Judith, and was told that she should not, as a member of the Secretariat, show an affinity to one region more than any other.

In reaction to the forceful direction of attention, the 'gaze' of Oxfam and the member groups, the members of the IS created other 'things' that side-step the 'gaze' vector (see Figure 8.4). Different members of the IS responded to the pressure of visibility—the attempt to make the purportedly opaque decision-making structure of the IS knowable and familiar—by trying to disappear in various ways from the terms of the game: if the Federation is based on national member groups, the IS was going to be composed of multiple nationalities, people who identified and presented themselves as transnational in their identity and experience. Thinking with vectors in fields of forces imagines the dance of relationships to include the effects of the magnitude of power and the directionality of different types of actions, self-presentations and self-identifications.

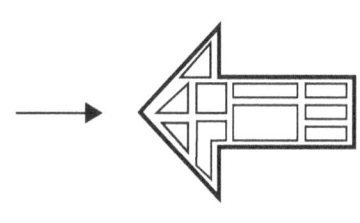

The activists at the FoEI IS shift the register of their actions in a way that is not recognised as powerful – by emphasising their multiple nationalities the activists of the IS are not understood as powerful along the lines of representatives of national member groups, which in FoEI is how votes and power are primarily structured.

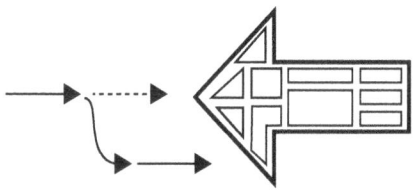

The actual actions of the IS are not changed but because of the change of register instead of being perceived as against the principles of decision-making of FoEI they are no longer as visible and in effect are forces against the flow of the FoE national members groups.

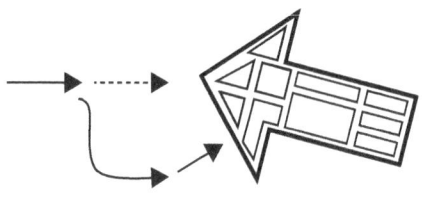

The effect is that the IS can make a difference to the direction of the FoEI activists direction.

Figure 8.4 Vectors changing field.

Note: By portraying the members of the International Secretariat as having multiple nationalities, as facilitators and so on, their power is not as easily visible.

Deploying FoEI

In FoE Brazil, not all the activists direct their attention to getting to know and having an effect on the Secretariat, or what is perceived as the thing '*foi*'. Andrea, coordinator of FoE Brazil, unlike Mariangela, knows the IS, and many people in FoEI, albeit not in a discursive sense. She strategically deploys and maintains the opacity of FoEI, and therefore the effectiveness of FoEI as a structure, as a thing. Similar to Stine, Andrea puts to work the image of FoEI as a solid entity because familiarity has made her aware of its fragility. Nonetheless, she deploys the notion of FoEI as solid in two separate ways. Both strategies stem from her perception that the continuation of FoE Brazil beyond the current set of volunteers, employees and activists needs protection. NAT needs to be secured.

Andrea actively deploys the perception of FoEI as solid towards audiences she imagines as external to both FoEI and NAT. One example concerns an organisation based in Brazil called 'Friends of the Earth Amazonia Brasileira'. As I described in a previous chapter, this organisation was, until 2007, an affiliate member of FoEI. It was set up by an Italian man, Roberto Smeraldi, who was originally a FoE Italy activist. The projects his organisation manages in the Amazon include a radio network for various indigenous communities to use for communication and to report illegal hunting, and more recently, to work with those indigenous communities to produce sustainable forest goods for international resale that would bring income and visibility to the communities. These communities are threatened, among other things, by encroaching cattle farming, illegal logging, the state's plans for more roads and infrastructure to pass through the Amazon and the strict conservation laws that often prevent the inhabitants from adapting their practices, making it difficult to respond to the changing availability of resources (see Adams 2003). Andrea welcomes any confusion in Brazil between FoE Brazil and Amigos da Terra Amazonia Brasileira: 'It gives the impression we are a big organisation', she laughed. But she meant it seriously.

Another example brings to mind Stine's concerns about funding. Andrea plays up the participation of FoE Brazil in the FoEI Federation when she makes presentations, has conversations or prepares documents for international funding bodies. In particular, FoE Brazil receives funding from the Heinrich Boll foundation. According to Andrea, being an active member of a recognised international federation increases the perceived competence of FoE Brazil. Upping the stability (perceived as credibility) of FoEI also increases the credibility of FoE Brazil and helps to secure repeat funding. Andrea travels to Germany almost every year for meetings at the Heinrich Boll foundation head offices where she reports on the year's work and presents proposals for funding for the following year. While in Europe—in 2006–07 it was her third visit—Andrea takes every opportunity to present at public conferences or seminars. She makes available documents that portray FoE Brazil as the Brazilian member of FoEI. But notably, she has never included in her month-long trips to Europe a visit to the IS in Amsterdam. The reason she gave me was that she does not like the way collaborative work actually turns out in FoEI campaigns and working groups.

Andrea says she knows the 'Federation' very well. When she describes instances of 'knowing the Federation' she mentions the number of people she is acquainted with, and those of them who are her actual friends, and the work of other people that she either knows through firsthand collaboration or through the related experiences of her friends and acquaintances in FoEI. She does not like to dedicate too much energy to FoEI because of the lack of work that actually gets done and the conflicts in the campaign groups. Andrea thinks that the work being done by FoE Brazil in nation-wide umbrella cooperations—such as the GT Energia that she coordinates, and the Rede Brazil that Mariangela coordinates—are more effective, and a better use of energy and money.

She explained to me how particular people she worked with in a project with FoE Paraguay have in the past applied for funds and regional projects, but then never delivered on the work. She described the harsh fights she has heard about in the FoEI campaign groups, that lead people to leave the groups, and how there is the same 'shit' in the climate groups as much as in the 'International Financial Institutions' group (a FoEI campaign group). She tells of how FoEI Amazonia Brasileira was pushed out because Isaac from Costa Rica does not agree with forest certification at all, and does not want a successful project of the sort in his area (Central America). Whereas Andrea admires the work of the founder of Amazonia Brasileira, I have not heard her talking about these things with others in the office. We talked about them in her home, or on a pickup van far out of Porto Velho in the northeast of Brazil, or in the car—but never in the office.

On the one hand, Andrea deploys the image of FoEI as a coherent and transcendent thing, because in this form it contributes to bringing the effect of further funding. However, she does not perceive FoEI as this coherent thing, since through her familiarity and knowledge of people and goings-on the shifting and human constitution of FoEI is transparent to her. Unlike for Stine, Andrea sees its fragility as a weakness, something to be not protected but avoided. On the other hand, Andrea makes a strategic effort to maintain the 'coherent' feel of FoEI in the eyes of the members of FoE Brazil. In the office she does not encourage participating in FoEI fora (which are anyway mostly in English, and therefore not really accessible to any FoE Brazil activist except herself). However, at NAT's annual retreat, which lasts for two days during which all the volunteers and employees get together to review the past year and plan the year ahead, a main *pauta* ('point' on the agenda) was 'FoEI Mission, Vision and Values'. Andrea insisted that this should be addressed, because it was part of their obligation as a FoEI member group. We spent at least two hours discussing each statement in detail and exploring resonances with the aims and actual practice of FoE Brazil as it currently is. Fulfilling obligations to FoEI is also strategic: three of the projects running while I was there included salaried positions funded by FoEI's Membership Support Funds.

There was no mention during the whole meeting that the Mission, Vision and Values was supposedly derived, at least in part from their work and then combined with that of all the other groups. Basically, they were comparing their own practical mission, vision and values with that of '*foi*' without being aware that their own work was part of the constitution of those statements. No one in FoE Brazil was aware of the SVPP, even though Noemi was in Abuja as a NAT representative.

At the international meeting, Noemi and I did not manage to communicate in English, and I had not yet learned Portuguese. In both the regional and the plenary meetings, Noemi needed someone to translate the Spanish into Portuguese for her. The language difference created a very real barrier in the extent to which Noemi could follow or participate. When I got to know her in Brazil she was not the shy, reticent person I had met in Abuja; she spoke eloquently and persuasively in Portuguese. Interestingly, however, although Noemi did not become familiar with the details of the processes at Abuja and how they related to the work done in Brazil and at the ATALC meetings she helps to organise, her understanding of FoEI as a thing is of a welcoming, supporting thing.

Learning Not to Question, Unprotected Backs

In 2005, Anthony managed to secure FoE Malta's participation in a capacity-building project, called Adelante, organised and funded by FoEE. Other activities of the project included two week-long meetings with representatives of all the FoE groups participating in the capacity-building project. David, Louisa, Elisa and Anthony were the activists who attended those meetings and dedicated the most time and energy to planning the project.

Once the mission, vision and values were created during that first retreat, the second action expected by Roger (the mentor) was to reduce the number of campaigns that FoE Malta worked on to one. At the time, FoE Malta had six campaigns: land use, waste, domestic chemicals, social justice, agriculture/GMOs, and water issues and energy (which included a number of projects ranging from educational and empowerment projects to research into the water and energy cost of tourism in Malta). At the time, FoE EWNI's largest and most successful campaign was the Big Ask. Each FoE EWNI local group (distributed in over two hundred localities in England, Wales and Northern Ireland) campaigned for local inhabitants to write to their local MP, asking why the percentages of emissions that contribute to climate change were not being reduced through year-on-year cuts. At the time, the UK had legislated that by 2020 it would reduce emissions of greenhouse gases by 20 percent. However, by not legislating on how much the percentage should be reduced per year, no practical action was actually being taken to enforce this legislation. The Big Ask campaign also enlisted Thom Yorke, lead singer of Radiohead, to front the campaign. Together with Tony Juniper, they organised Big Ask events where Yorke sang and spoke about the campaign.

One of the most important activities of the capacity-building project was that six FoE Malta activists got to spend a week in the offices of FoE EWNI, right in the middle of the Big Ask campaign. On their return Anthony, Louisa, David, Elisa and Mario could not stop talking about the impressive Big Ask campaign, the size of the FoE EWNI offices and the professionalism they encountered there. The Big Ask was going to become a FoE Europe campaign and FoE Malta, as with all the other FoE Europe groups, had the choice to join the campaign.

The next capacity-building task was a retreat organised specifically for the creation of a strategic plan for FoE Malta, which included the task required by Adelante to focus on one campaign. The outcome of that retreat, which all the active members of FoE Malta, except myself and one or two others attended, was to join the EU Big Ask campaign and focus entirely on it. From then on for the next three to five years, FoE Malta would focus primarily if not exclusively on climate change. I do not know how the decision on this was taken; however, I contested that decision with Anthony. I asked him about land issues, that apart from being urgent with the breakneck rate that the Maltese countryside is being transformed into buildings, have historically been FoE Malta's main area of success—as with the Verdala golf course and the Mnajdra landfills. Anthony responded by saying that FoE Malta does not have enough volunteers or employees to cover more than one campaign and that the active campaigners all want to work on climate change. I protested to both Anthony and Roger that as a member of FoEI, FoE Malta cannot be a single-issue organisation, and that capacity-building funds were given to FoE Malta on the basis of its being a FoE member. The result is that FoE Malta is presented as having seven campaigns, all of the above plus climate change. The campaigns are still present on the website; however the new activities embarked on all circle around climate change. Of course, Anthony reassures me that were volunteers to take the initiative to organise things within these other campaigns, there would be no barriers to that. In fact, Jurgen, Anthony, Louisa, David and Elisa still attend the Nature Group meetings and events regarding land use. They do not, however, take leading roles in the organisation and running of activities or events planned there.

Let us return to vectors. At the time, David, Louisa and Elisa, unlike Anthony, had never met activists from other FoE groups; they had never attended any of the General Meetings of FoEE or of MEDNET. Their only experience of FoEE was through educational meetings with FoE EWNI, and exchange/educational meetings with the other Adelante participants. Before the project, both FoEE and FoEI were little known entities; to Louisa, David and Elisa they were opaque entities. When they gained familiarity with other FoEE groups, they did so through 'educational' meetings, unlike the newcomers to General Meetings, where every representative has an equal vote, that we met in Chapter 6.

During the capacity-building meetings the 'experienced' twin was positioned as the more powerful FoE group: more powerful in terms of numbers of activists (volunteers and employees), more experienced in terms of size, width and type of campaigns, actions and projects, more successful as an ENGO. The situation in which the FoE Malta activists got to know other 'FoE' groups did not lessen their sense that FoE EWNI and FoEE (not to mention FoEI) were powerful 'things'. So much was this that the process of the Adelante capacity-building project was not questioned and the actions expected of FoE Malta by the 'experienced' twin, FoE EWNI, through

Roger, were all carried out as far as possible—even at the expense of Elisa's spending more than forty hours a week of work when she was only paid for twenty hours. Elisa said she 'had to do it', or 'who else would?'

In terms of vectors, one's education of attention can bring with it other types of opacity—some vectors can cancel others out. David, Elisa and Louisa directed their attention to the specific meetings, held as part of the Adelante project, in order to get to know what was being presented to them: the Big Ask campaign and the 'experience' of the twin. This experience cancelled out their focus on the expertise gathered by FoE Malta (in the work of Izabela, Jurgen and Anthony) with regard to land use and agriculture issues. These processes of learning in Louisa, David and Elisa's knowledge of FoE Malta cancelled out knowledge of FoE Malta's previous ability to function as a significant environmental public lobby group without receiving any funds from government or corporate sources. By keeping itself small and 'unprofessionalised', FoE Malta had managed to sustain its activities without pressure from the government.[2] Furthermore, it cancelled out the decision-making that is particularly attentive to process—questioning the process of capacity building was not automatically part of the learning experience, as it is in the FoE GMs. Through this focus on the Big Ask and climate change, FoE Malta has become more disciplined and has participated in more events in the past two years than it had in the previous five. My point here is not to criticise the work done through and after the capacity-building project, but simply to emphasise how different contexts of gaining familiarity can produce different understandings of what can be changed, and of where a person's energy and agency can be directed. The following description of newcomers at the FoEI international meetings is an example of how, when 'process' is the focus of an education of attention, newcomers learn very fast to attend to decision-making processes and to question those processes.

At the inter-BGM in Swaziland, I participated in the Membership Development workshop, which was to take up three days out of the five-day meeting. The workshop focused on capacity building for FoEI, asking how the members present (as representatives of their respective national member groups) would choose to define it. For the first two-hour session of the workshop, Roger—who was the coordinator of the Adelante project and FoE Malta's 'experienced' twin—presented himself and his experience with capacity-building and membership development and then presented his plan for the activities of the workshop. The first day was to include small teams working on what they imagined a 'strong network' to be, feeding this back to the group and as a group coming to a consensual description of a 'strong network', then breaking off into smaller groups to devise actions to reach that imagined 'strong network'—again coming back to the whole group to feed it back and agree on a plan and on who would take responsibility for what.

After Roger's presentation, rather than dividing off immediately into groups, many experienced FoEI activists questioned the process that Roger had offered. They asked by what authority he was proposing the plan, who had chosen him to facilitate the meeting, was there not going to be space to change the process? These questions were directed to Griet, the director of the IS. Her reaction was first to defend Roger, explaining his extensive background in capacity building for FoE EWNI, then the great success of Adelante and the previous FoEE capacity-building project. After that, she turned the questions back to those who posed them and, by catching the eyes of other people around the room, to the whole group. She asked them what they expected from the workshop. At this point, the discussion became polarised between those who wanted to have specific outcomes and a clear plan of action to follow, and those who wanted to use this space for open discussion but to delay making a plan of action. Earlier at the IS, when the Secretariat was facilitating the negotiation of the agenda for Swaziland, Imee reflected that the European groups are mostly the ones who want the outcome of the international meetings to be decisions on actions, formulated plans with people responsible for this or that to happen. On the other hand, after the crisis with FoE Ecuador, the Latin Americans wanted to have more space to discuss ideological differences among the different groups. She hoped that the Swaziland meeting would not be caught between these contrary expectations. She hoped that the agenda, and the facilitators they had found, would manage to find a balance and that the 'groups' would not be frustrated.

With the entire morning dedicated to discussing the process of the workshop, a discussion which was not included in the plan that Roger had presented, it was agreed that we would carry out the first task—small groups discussing what a 'strong network' means to them. Then the following tasks would be carried out, but with someone from the IS assisting Roger in facilitating the group. There were a number of activists in the group, less experienced with FoEI meetings, who did not speak a lot in the morning discussion—among them was a young woman, Cynthia, who had worked with FoE Australia for three years but had never come to a FoEI international meeting. She expressed her frustration at the morning's meeting: 'FoEI meetings are so strange! We spent more time discussing process than anything else!' By the end of the week, this same young woman had facilitated a number of group sessions as well as the most contentious group meeting held in the evening about the then upcoming climate meeting of the UN in Bali. At the end of the workshop, when people were thanking the organisers and the IS, the main comment coming from experienced FoEI activists was how well Cynthia had facilitated, making space for everyone even when their positions were so different. Cynthia's learning was directed towards facilitation of a particular kind, where this does not mean presenting ready-made proposals (as Roger did), but being the person to ensure that the proposals made by the group did not include one voice to the exclusion of all others. Here the context of Cynthia's becoming familiar with FoEI was a situation in which all members were considered equally experienced, and correspondingly she learned not only how to be included in the active making of the 'process', but how to lead the meetings herself. The creation of space within the FoEI international meetings, where representatives are allowed to question process, allows for the imagination of having 'structural' power, or in other words the power to actually change the 'structure' of FoEI decision-making processes. The feedback loop that comes from this is different from the way of learning that maintains a hierarchy between experienced groups and capacity-needing groups as in the Adelante project.

Both situations, the capacity-building meetings and the GMs, are instances of where FoE activists have learned about FoEI. However, by taking into account what vectors were in play we can see how in the latter situation, at the GM, FoEI might remain an entity but it is not opaque, and is therefore malleable. During the Adelante meetings, however, the activists' attention was not directed towards the working processes of their experienced twins. Thus the experienced FoE group remains opaque and 'powerful' to the extent that much of the experience that 'FoE Malta' does have was replaced by the new work suggested by the twin.

Being Aware of Unprotected Backs

Another aspect of the metaphor of fields of forces and vectors is that they are dynamic and constantly in flux. Anthony's experience of working with a para-statal organisation in Malta shows how direction of attention needs to be sustained; familiarity is not maintained automatically and as transparency can be learned, so also a certain opacity in practice can also return. In 2000, a couple of years after having completed a diploma and an advanced diploma in computer science run by MITTS (Malta Information Technology and Training Services), Anthony found a job at this institute. MITTS provides the IT support to most government offices. Anthony was supporting the implementation of these software packages at the AFM (Armed Forces of Malta), Health Department (immunisation department) and Parliament (software to handle parliamentary questions). At the time, he was already a core activist with FoE Malta. Talking about his experience of working with the Maltese government he said, 'I learned a lot about the government working at MITTS, who to go to, how things work and don't work and the general mentality. I use this knowledge when dealing with people [for FoE Malta campaigns] but of course things change, people change—[lose] contacts mainly.' In 2001, Anthony left MITTS to work at the Maltese branch of the International Ocean Institute, where he designs data analysis software, their brochures and website. In those years, he went from being a core member to being the coordinator of FoE Malta, and all along mobilised his contacts in 'the government' to, for instance, find information, lobby for support or find the right person to lobby and send information to.

The 'government' is not an opaque Leviathan for Anthony. The knowledge that the government is accessible through human relations is a constant reason for hope that FoE Malta's work is not impossible. It is a hope that continues even in the face of constant setbacks, as the FoE Malta activists perceive them: MEPA (Malta Environment and Planning Authority) continually grants permits for construction in areas outside of the development zones; MEPA Boards continue to grant permits against the advice of the reports compiled by their own officers; six years after EU accession the government has not produced a policy-paper for safeguarding freshwater; hunting continues to decimate migratory birds on their way to North Africa; Malta ranks worst in a survey of climate change indices; the government still has not produced policy papers regarding alternative energy, and there are only minor incentives for alternative energy use; there is even less enforcement against over-use by industry; new roads are regularly proposed, some built, taking up countryside or agricultural land, encouraging increased personal vehicles and the urban sprawl; rent laws have as yet not been revised, leaving approximately a quarter of homes in Malta empty, while new houses are continually being built, and so on.

Anthony's knowledge of the 'government' allows him to do something about the actions of the government. Simultaneously, his knowledge of the relationships that constitute an otherwise powerful institution, allows him to turn his hopes, at least partially, into reality through his work with FoE Malta. Finally, Anthony's experience also shows how attention can be directed in more than one direction, but not in all directions simultaneously. So when Anthony left MITTS and started remunerated work in a new organisation he needed to direct his attention to the relationships entailed, and could not keep up his old relations 'in government' as much as when he worked with MITTS.

Conclusion

To sum up, I suggest that the notions of vectors and fields of forces are useful to think about power and powerlessness, and to account for shifts in power. First, the notions are ethnographically precise. FoE activists experienced situations in which they had 'agency', but also situations in which they felt powerless or where 'things' were so opaque that questioning them was not even possible. Second, as analytical tools vectors and force fields also allow for connections among symbolic, material, political and apparently not political (for instance educational) spheres to be recognised. When the members of the Secretariat emphasised their 'facilitator' role and transnational identity, they were successful in allowing them to carry on as an office (thus the Secretariat was not disbanded, FoEI did not always have an organised Secretariat), and not because they had won a zero-sum game of power 'against' those FoE groups who accused them of being aligned with European environmentalist styles. The leaders of the Secretariat want to be effective in maintaining and increasing the capacity of the Federation to 'hang together'—they are more interested in influencing the types of decision-making processes that go towards making 'FoEI' positions. In other words, the core Secretariat members are directing their attention towards influencing 'FoEI', and their emphasis on facilitation is a shift into a different register of action. While facilitation and transnationality are not the 'unpowerful', 'unpolitical', 'non-aligned' positions they are supposed to be, they comprise a field of force that does not interfere with the struggles of environmentalism that triggered the accusations towards the IS in the first place.

Vectors and fields of force 'explain' how things like 'offices', 'institutions' and other 'structures' actually do have effects even though people still have and exact agency. My understanding of structure and agency builds on Giddens's theory of structuration but extends it to include intentions and an ethnographic description of how 'unintended consequences' that give power to structures can change the power of those structures through the 'direction of attention'.

The effectiveness of structures may take different forms. People perceive that if they direct their attention to structure *they* will have no effect—no agency. These situations may either arise because they do not have the necessary relationships, as with Mariangela's efforts, or Andrea's decision not to engage with FoEI work, or Anthony's reflection that he no longer has the right contacts, the knowledge needed to crack the 'thing'. Or, having agency towards a particular 'thing' may not even be considered as a possibility, so that no attempts are made to direct attention towards it. This is what I call unprotected backs, and the experience of Diana, Louisa, David and Elisa are examples. Finally, to return to the example I began with, Izabela started off with an unprotected back towards 'structures', but when vectors are compounded—that is, when her direction of attention was compounded with the direction of attention of the rest of her colleagues—the fields of force changed accordingly.

Notes

1 A 'thing' itself, which also deserves exploration, which however is beyond the scope of this particular study. However, see below for a description of the FoEI IS office.
2 Contrast this with the case of Nature Trust Malta. The main drive of the ENGO in 2000 was to 'professionalise' the organisation (Gatt 2001). In 2006, Nature Trust participated in the first of two of the largest environmentalist protests there have been in Malta to date, when all the members of the Nature Group called the members and anyone else to join the march in protest against the extension of the development zones. A week after this protest, Nature Trust and DLH (also closely involved with different patronage relations with the state and Nationalist party, Boissevain and Gatt 2011) publicly retracted their criticism of the government's choice to extend the development zones. Rumours through the nature group emails ran that Malta's prime minister at the time, Lawrence Gonzi, had personally called the president of NT threatening to cut the funding (EU funding) for their reserves and to cancel their agreement on the new nature reserves.

9 Experiments with Political Form and Process

In the spirit of the anthropological 'sideways' glance (Ingold 2007) and seeking the 'actor's point of view', the chapters so far have been attempts to understand the position and experiences of FoE activists. Nonetheless, throughout critical reflexivity (Okely 1996) and systematic doubt (Milton 1993: 6) were necessary to analyse the activists' experiences and practices. However, in this final chapter, the stance is shifted to 'critique' (Marcus and Fischer 1986). In this chapter, I highlight various apparent mismatches between what is said and what is done in FoEI, specifically around the notions of flatness, decentralisation, inclusion, consensus and facilitation.

FoEI structures are described as 'flat', yet there are often hidden hierarchies. As a federation, FoEI aims to have a decentralised structure, yet certain practices centralise important decisions. Inclusion is central to the FoEI 'Mission', yet there are significant forms of exclusion embedded in being an organisation. Consensus is often used as the decision-making tool to obtain inclusion, yet it too can become exclusive when used by individuals to 'veto' collective decisions. 'Facilitation' is preferred to 'leadership' in many FoE contexts. Many FoE activists imagine a facilitator's role as consisting primarily in the obligation to serve the group, rather than have the group serve a leader's desires or interests. It is commonly imagined among FoE activists that while centralised organisations have leaders, chief executive officers, or presidents, decentralised federations have facilitators and chairpersons. However, in important FoE contexts, the pivotal and powerful position of the facilitator is not a point of focus; rather the attention of FoE activists is directed elsewhere. In these situations, the difference between (decentralising) facilitators and (centralising) leaders is blurred.

I explore these notions through the lens of the Strategic Vision and Planning Process (SVPP) and related activities. I chose this focus for two reasons. First, the SVPP was pivotal to the federation's restructuring to be more inclusive of global social and cultural difference. It emerged from a crisis when FoEI as a 'worldwide' federation was threatened. The SVPP was the first reflexive and deliberate attempt to make FoEI's decision-making processes inclusive. Previously, 'inclusiveness' was associated simply with having member groups from all around the world. Second, since the SVPP is talked about as specifically an experiment in decentralisation and inclusion, and in the achievement of flatness and consensus, it is also the context in which the opposites of these are most apparent.

The activists are highly reflexive and monitor both their own actions and those of others, as well as the situations around them. Analysing the mismatch between discourse and practice in this context needs to take this into account. At times, the activists are fully aware that there is a 'mismatch' between rhetoric and action. Consider the following extract from my field notes.

2 October 2006—third day of the BGM in Abuja
I was just talking to Henry from Ghana, we had been sitting together for most of the inner-outer circle discussion of the 'strategic maps'. Carlos had asked to include the word 'political' into one of the bubbles. The bubble was approved. Henry said, 'look how they use political, everything is political to them', referring to 'South American' FoE groups.

Henry—The words are so important for them, they are full of their experiences, so when they talk they are not being bitter, or pushing, but it is because that is how their everyday life is, everything is political to them. Some people seem to think that words such as 'radical' were too strong, and they could not understand why the Latinos were pushing so hard. But the word 'political' has a special meaning to them. It comes from their experience, every day they struggle, in everything they do.

Caroline—Why do you think that happens?

Henry—The problems they deal with mainly in Latin America, have to do with the big corporations that come and take over the land, for instance a large plantation of timber, and they will use the military to protect it, taking away any control from the indigenous people . . . most of these corporations are from the United States, and they become citizens, they live there and they influence the government, they also keep really strong contacts with the United States, so it feels like there are foreigners who live among them who are taking their land and their resources. Here in Africa we don't see it as someone from the outside coming in—it is not recognised at least, things will be blamed on the local leader's capacity [or lack of it] to promote his community's needs. On the other hand, in Latin America there is still the feeling of colonialism, that the corporations come in, live on the land and take the resources for the wealth of the North, where they still live, even though they become citizens.

Henry (again)—We meet once a year in campaign meetings, this year we are meeting twice because of the BGM, and by working with them I can really understand why they are so strong about 'politics'. We need to give them space, because even in their own country they are pushed and squashed by the experiences they have, by their realities, so they feel they need to fight for space here too and we need to make them feel that they have space.

Caroline—Do you think they have enough space in FoEI?

Henry—Yes, yes, but I wanted to say that when they talk during these meetings we need to understand them, I understand how important politics is to them—everything is political: membership is political, finances are political.

Henry is a member of the FoEI ExCom, and from this conversation it is clear that he has reflected on the differences in experience and expression among the different 'regions' of FoE groups, especially those from the Latin American FoE groups. The extract shows how FoEI activists' reflexive thinking is comparative and how what might appear as a mismatch between words and action needs to take into account the activists' own reflexive critique.

I am reminded, by the approach of FoE activists, that to write a chapter about the difference between what FoE activists say they do and what they actually do necessarily means choosing one definition of these words over other possible ones, as well as one interpretation of these actions out of other possible interpretations. However, I choose to make those judgments. In this chapter, I specify from where I derive the definitions of concepts and words I critique and for whom my interpretation of actions would be appropriate. Nonetheless, I recognise that I am inevitably putting forward my own framework for interpretation and judgement. Well aware that our critiques as anthropologists may compromise the strategies deployed by the communities to which anthropology purports to give a voice (Brosius 1999: 280–281), this chapter offers a constructive criticism. Admittedly, as a FoE activist in the past myself, I share the hope of bringing about 'environmental justice'. My hope is that this critique may also be of use to FoE activists themselves.

I begin with a reminder of the SVPP (see introduction), its aims, and the processes and activities that constitute it. Stemming from this I describe five ideal notions as they are brought to life in FoEI discourse. For each notion— flatness, decentralisation, inclusion, consensus and facilitation—I present apparent contradictions between ideals and practice.

The Strategic Vision and Planning Process

The Strategic Vision and Planning Process (SVPP) is a plan for activities on which FoEI embarked on in 2004. It was planned to last at least ten years. The aim of the SVPP (as it is described by Anthony, members of the Secretariat and the ExCom and FoEI documents) was to carry out specific tasks that would create fairness in the decision-making processes of FoEI. The SVPP was adopted by the EGM in 2003 as a way to address the issues raised by the crisis in 2002 when the Ecuadorian FoE group left the network. The main complaint of the Ecuadorian group, a number of South and Central American FoE groups, and a couple of European and Eastern European groups was that the decisions and public communications of FoEI favoured the interests, or approaches to environmentalism, of 'European'

FoE groups. Although FoEI was set up as a federation and the official FoEI documents described it as a 'grassroots federation', during the crisis many groups argued that it was not 'truly' a federation of equal groups.

An EGM in Colombia approved the SVPP, and a coordinating team was set up. The team was called the Strategic Planning Transition Team (SPTT). The team was composed of members of the ExCom, members of the Secretariat, the programme coordinators and a few other FoE activists that the ExCom co-opted for their experience. Finally, the EGM also approved that Vera, a professional facilitator, should be employed to develop the SVPP tasks and manage the process.

Between FoE Ecuador leaving in 2002 and the EGM in 2003, the ExCom and the Secretariat had developed proposals for the SVPP, identified a process manager who would be acceptable to the different groups, *and* found funding (Oxfam Novib). The EGM was presented with concrete proposals, although Anthony describes the discussions at the EGM as long and painful. However, as the SVPP progressed, the representatives at the SVPP workshops (both experienced FoEI activists and newcomers) became more and more knowledgeable and courageous about questioning the whole process—from proposals for the tasks to be carried out during the workshop to the ways decisions would be taken.

Before the SVPP, not much attention had been paid to the actual processes of how decisions were taken in FoEI. The initial spirit of the federation was deliberately anti-organisational. Who could join the federation, and how these groups chose to work, would not be dictated by 'some tired formula from the top'. The lack of bureaucracy was an explicit move on the part of the founding groups.

As the federation grew, however, some elements of structure were proposed and discussed: having a general meeting and a statute, an Executive Committee (1983), a Secretariat, first rotating and then settling in Amsterdam (1985), and campaign groups, were discussed in detail at the time when they were set up. For instance, FoEI came to have organised 'campaigns' with campaign coordinators and campaign groups in 1985. The structure and content of these 'campaigns' was the main point on the agenda of the General Meeting in Scotland that year. The structure and character of the ExCom was the focus of discussions at another FoEI General Meeting, where at first this committee was primarily set up to organise the annual meetings from year to year and gradually in subsequent General Meetings was given more and different sorts of responsibilities (Link Archives, Amsterdam). In an interview, the former chair of FoEI, John Hontelez, explained that at first the members of the ExCom acted as representatives of their national organisations. After some years, a General Meeting agreed that it was necessary for members of the ExCom to be elected not primarily as representatives of their national member groups, and that the responsibility of the members of the ExCom was to consider the federation as a whole. Thus, even the relationship between ExCom members and their respective national FoE groups was discussed and given attention.

Finally, the character of the General Meeting changed through discussion at the meetings themselves. The General Meeting began as an informal annual gathering of environmentalists, often at big international events where activists would be primarily present for the event. In 1983, the annual meeting became the 'General Meeting' and became instituted as the 'highest decision-making organ' of FoEI. In 1998 it was proposed, discussed and agreed that the annual meeting would become a biennial meeting.

One of the results of the discussion at the 2003 EGM, which analysed the causes of the FoE Ecuador crisis, was the reflection that during General Meetings discussion was focused on 'actions': what campaigns the federation should launch, which meetings to attend, and so on. The focus on 'actions' left no space to discuss underlying principles, including different approaches and understandings of politics and environmentalism of the FoE groups (SVPP working document 2006: 15). Previous General Meetings had even discussed this, and the possibility of discussing understandings, beliefs and ideology was rejected in the belief that these discussions would slow down the network's campaigns (*ibid*: 15–16). Due to the crisis, it was considered necessary to discuss beliefs and principles and organise decision-making processes.

At the time of its foundation, FoE activists believed that inclusiveness would arise from a *lack* of structure, and a lack of specific discussion of principles. Now however, as the SVPP shows, FoE activists are experimenting with organised reflexive discussions to achieve 'inclusiveness'. Whereas before, what was being included was representation from a variety of countries, now, as we can see in Henry's reflections cited previously, 'inclusion' refers to the FoE representatives' personal understandings and positions.

The SVPP had three phases. The first phase developed the federation's 'Mission, Vision and Values'. The second phase developed a ten-year strategic plan for the Federation. The final and longest phase, which in some ways is still in progress today, is the implementation of the strategic plan. Each of the phases began with a workshop at the annual meetings of the FoEI regions that had been prepared and planned in advance by the SPTT.

The workshops were facilitated by a member of the SPTT and observed (often through participant observation) by one or more members of the Secretariat on the SPTT. In the European regional meeting, the SVPP workshop was run over two days just before the annual General Meeting of this grouping, which has a formal organisation. The ATALC, the African region, the North American region and the Asian-Pacific region did not have formal regional groups before the SVPP, but the FoE groups from within these geographical regions came to have annual meetings through SVPP funds. Now the FoEI website shows the different national FoE groups within these 'regions' or on 'the world map'. Prior to the SVPP, the groups were only marked as dots on a flattened world map (at the time with Europe at its centre, even this has now been changed—website accessed 20th April 2016). The rhetoric is that change in the way the groups are presented on the website is also part of the SVPP actions taken to equalise the different 'regions' (on the website as in FoEI decision-making). 'Regional balance', having a similar number of FoE activists from the different FoEI regions, became strictly applied as a result of the SVPP, in the composition of campaign and programme teams, although it had long been observed in the composition of the ExCom. The countries where the BGMs are hosted also follow this idea of balance. If in the early years of FoEI most of the annual meetings were held in European countries, now, to create a balance, most of the General Meetings would be held in other FoEI regions.

After the five regional meetings in 2006, the phase of the SVPP dedicated to creating a strategic plan, the members of the SPTT collected all the 'flip chart papers' with the final outcomes of the regional discussions. These documents were then analysed by the SPTT. The five regional strategic plans were compiled into a single document that was understood to, and hoped would, incorporate *all* the work of the different regions. I did

Figure 9.1 Reviewing the outcomes of discussion of other groups at the FoEI BGM in Nigeria, 2006.

Note: The regional discussions of vision for FoEI and the timeline, as well as the ten-year strategy negotiations shown here, included both constructive production—where people made suggestions (these were simply added to a list on flip charts where everyone could review them)—and debate, where people disagreed or persuaded others to include or remove a point. Like the meetings in Estonia described in Chapter 6, the flip charts acted as 'boundary objects' that allowed the participants to check whether they agreed with each other. They were also the 'things' they were making—these were the preliminary documents, they then became the humus for further production of different documents.

not have access to the work of the SPTT at the time of my fieldwork, so I cannot describe the process used to compile these different documents. Finally, the document was presented at the following BGM, in Abuja, to be discussed, elaborated and then voted on. The documents presented were strategic maps: an overall strategic map, and second-level strategic maps that elaborated each strategy 'bubble' from the overall map (see Figure 9.1).

Like the small breakout groups in Estonia that I described in previous chapters, the regional groups were intended to create space for more representatives to speak and to contribute and therefore to be 'included' in the creation of the documents that were to become the defining documents of a more equitable FoEI. In the regional groups, it was assumed that regional representatives would be more at ease, with fewer language barriers.

Simultaneous interpretation at the FoEI GMs is provided for Spanish and French (see Figure 9.2). During the Swaziland inter-BGM, it was decided that the default language for the international meetings would no longer be English, and that a future GM would be held in Spanish with English (and French) interpreters. Simultaneous interpretation is one of the most obvious tools used to include representatives from different parts of the world. In

Figure 9.2 The mission and vision projected onto the wall in English and Spanish, in the process of being discussed during the FoEI BGM in Malaysia, 2004.

Source: Photo by the FoEI IS

the FoEE AGM in Hungary and in Abuja, the Georgian and the Bulgarian representatives were virtually left out of all discussions because they understood and spoke neither English, Spanish nor French.

As can be seen from the photo, the flip charts from the regional or smaller groupings (in the case of the FoEE AGM in Estonia and in Hungary) are hung up around the main meeting room. Participants are encouraged to visit the different charts to get to know them, and to prepare for the subsequent discussion where these different points will be incorporated into a single document. This second step of presenting or displaying before the discussion on incorporation begins is very important. It creates the expectation that during the 'deciding' discussion that follows, participants will consider the points raised by all the groups, and not only their own.

This second step marks a shift when participants move from thinking about their own interests and experiences (and those of their regions) to thinking about 'FoEI' as a 'whole'. There are variations to this 'display and get to know' activity. In Abuja in 2006, a 'marketplace' was set up. In each 'market stall' different strategic 'bubbles' were discussed, facilitated by an ExCom member of SPTT. Participants could roam or choose to stay with one discussion.[1] The results of these discussions were then hung up around the main room and participants were encouraged to examine the points arising from the discussion of other 'market stalls' (see Figure 9.3).

In Swaziland in 2007, however, the 'marketplace' was something else. For the first time at an inter-BGM, participant groups were invited to present the work of the national group on a table. Participants had an entire morning to visit the different 'stalls': tables with documents, publications, videos on laptops, stickers, posters and so on. The idea was to get to know the work of the other FoE groups. In Swaziland, also for the first time, every day of the inter-BGM was opened with a 'mystica'. Each region was asked in advance to prepare an activity, a presentation, or something that would give a sense of that region.

The ATALC region went first. In the plenary room of the meeting, with low lights, in a single file, about twenty activists walked into the centre of the room. All the other activists sat in a horseshoe shape around them, watching and wondering what a 'mystica' was. The single file of South and Central American activists became a tight knot and while they circled they placed green- and yellow-coloured pieces of card on the ground. The knot slowly dispersed and they left behind them a large collage of a cob of sweetcorn (see Figure 9.4). The 'mysticas' were an experiment, a space in the international meetings for groups to present themselves other than through discussion, something that would not otherwise occur in their highly discursive exchanges. The hope is that 'getting to know' the other FoEI groups will increase understanding. The processes of the SVPP not only increased FoEI time dedicated to reflexively discussing principles, ideologies and approaches, but also to sharing and exchanging knowledge that was not necessarily discursively based.

Figure 9.3 The FoEI 'marketplace' at the BGM in Nigeria, 2006. Developing the FoEI ten-year strategy through forms of discussion and decision-making derived from Open Space Technology.

Figure 9.4 The 'Mystica' performed by the ATALC regional group.

Flatness

Our Vision and Mission

Our vision is of a peaceful and sustainable world based on societies living in harmony with nature.

We envision a society of interdependent people living in dignity, wholeness and fulfilment in which equity and human and peoples' rights are realised.

This will be a society built upon peoples' sovereignty and participation. It will be founded on social, economic, gender and environmental justice and free from all forms of domination and exploitation, such as neo-liberalism, corporate globalization, neo-colonialism and militarism.

We believe that our children's future will be better because of what we do.

Our Mission

1   To collectively ensure environmental and social justice, human dignity, and respect for human rights and peoples' rights so as to secure sustainable societies.

2   To halt and reverse environmental degradation and depletion of natural resources, nurture the Earth's ecological and cultural diversity, and secure sustainable livelihoods.

3   To secure the empowerment of indigenous peoples, local communities, women, groups and individuals, and to ensure public participation in decision-making.

4   To bring about transformation towards sustainability and equity between and within societies with creative approaches and solutions.

5   To engage in vibrant campaigns, raise awareness, mobilise people and build alliances with diverse movements, linking grassroots, national and global struggles.

6   To inspire one another and to harness, strengthen and complement each other's capacities, living the change we wish to see and working together in solidarity.

FoEI Mission and Vision, 2004[2]

The most fundamental notion in the written constitution of Friends of the Earth International is that of being a federation. FoEI was originally set up as a loose network of environmentalists and both the looseness and the idea of equality inherent in a federation are encapsulated in the image of a flat structure, as opposed to a hierarchical one. Flatness, equality and freedom are central to the FoEI's mission of ensuring 'environmental and social justice'. In the mission statement, we can see how the ideas of flatness and of networks are associated with justice and equality: 'We envision a society of *interdependent people* living in dignity, wholeness and fulfillment in which *equity* and human and peoples' rights are realised'. To achieve interdependence and equity what is needed, according to the FoEI's 'Mission and Vision' is inclusion. Inclusion, in turn, is to be achieved through 'participation': 'to ensure public participation in decision-making'.

Importantly, the 'Mission and Vision' is specifically a statement of what FoE activists *strive* to bring about through their work. The creation of a participatory decision-making process through the SVPP is part of the experimentation in FoE activists' work. The creation of the Mission and Vision document through the SVPP was one of the ways in which FoE activists began realizing this aspect of their mission. In the notion of a federation, and in the effort dedicated to continually revising the work practices of the 'network', this collectively created mission is applied in experimental practice. The words 'network' and 'federation' are used interchangeably by activists. Both words are understood to connote structures that facilitate 'public participation in decision-making'. The same goes for the notion that FoEI is a 'grassroots' federation. FoEI chooses to work with 'grassroots' groups and communities to try and bring access to decision-making power where it is perceived to be lacking.

By contrast to a hierarchy, commonly depicted as pyramid-like, a network is imagined to be *flat*. Flatness in the discourse of many FoE activists becomes a gauge for how close something is to the ideal of equitable decision-making. Even the notion of 'regional balance' implies flatness. When a seesaw is balanced its plank lies flat on its pivot. If one region has more representatives, making the group *un*balanced, one side of the seesaw, a region, is higher than the other, or, to mix metaphors, has more weight. An extract from my field notes points to how 'flatness' is equated with the ideology of FoEI as a 'network'. However, the assumed flatness of the network (since a network is not necessarily flat), is a function of its opposition to the pyramid, which has to stand up.

> Although I had not attended a FoEE general meeting before, every time Sharbil spoke during the plenary discussions, there was a tangible tension in the air. Some people sitting on the sides of the large room, beyond the rectangle of tables and chairs that was laid out in the centre, spoke to each other in only slightly hushed voices, or focussed more intensely on their laptops. Sharbil, Roger explained to me later, is the new executive director of the FoEE office, he is an employee, his is not an elected post. A few months from then his performance as director was to be assessed by the ExCom (all posts elected annually by the general meeting). All 'campaigns' are proposed, discussed and agreed upon by the general meeting, Ian explained. Ian is an FoEE office staff member, he told me that Sharbil had decided to call many small-scale actions 'campaigns', without consulting the GM or even the ExCom. Sharbil had not even discussed this with his staff: 'suddenly we found we had all these campaigns! But many of the campaign websites are empty, or have one press release'. Ian was considering leaving his employment with

FoEE. The previous and very popular director of FoEE was visibly less flustered than Ian. He said 'Sharbil comes from Greenpeace, where they have a very hierarchical structure. The directors decide what campaigns to carry out and the activists implement them. Sharbil is not used to our way of doing things. We discuss everything as a team before taking decisions, even our salary structure is the flattest in Brussels, with the smallest difference between levels of employees, and we have the lowest salaries. This is really in the spirit of the network. We'll just have to see how he gets on'.

> 12th June 2006, Tartu, Estonia, first day of the FoEE AGM

The extract shows how flatness is opposed to hierarchy and centralisation. What ideal 'flatness' should be is unclear, however, especially in the light of the second part of FoEI's mission clause 2: to 'nurture the earth's ecological and cultural diversity'. No established form of decision-making currently exists, as yet, that fulfils the ideal of inclusion for human and ecological diversity. In his book *The Politics of Nature*, Latour (2003) proposes a new constitution, a manifesto, for how such a political process could work. There would be two houses, one in which matters of concern would raise their issues and a second house to institute (provisionally) matters of concern into essences (until externalities 'knock' at the door of the first house and essences return to being matters of concern). My contention here is that FoEI, and other environmental justice groups (see Di Chiro 1996), have been in the process of experimenting with this new constitution for some time. What Latour (2003) presents is more like a streamlined description of the work of environmental justice over the past thirty years. At the Abuja GM I voiced Latour's proposals, and argued, unsuccessfully, to the plenary that non-humans should be explicitly included in the notion of democracy. During the subsequent coffee break Otto, a long-standing member of the ExCom and of his member group (Switzerland), shared his thoughts on the matter with me. According to him, it was too early to use the phrasing I had suggested, due to FoEI's public. However, he said, in practice many FoE groups had been working for that for years, in the lawsuits that groups in Europe, Africa and South America had filed to give rights to different species and ecosystems.

FoEI is not alone in facing the challenge of reconciling diverse understandings and positions in policies. The notion of inclusion of human minorities ('diversity' in FoEI terms) is a constant matter of debate in countries like the UK. In Malta in 2010, the office of the prime minister was running an experiment with making policy decisions more inclusive by creating online fora for discussion of government policies. The greatest challenge in this project, according to the researcher running it, is how to reconcile the interests of different groups (minorities) in the creation of coherent policies (Derrick Pisani, Office of the Prime Minister Malta, pers comm.).

Consider the following phrases from the statements of the FoEI 'Vision and Mission' cited earlier: 'human and peoples' rights' from the Vision; and 'groups and individuals' from the third Mission clause. These statements were created at the end of many days of discussion at the Malaysia BGM in 2004. They include the insistence of a number of South American FoE activists, some of whom who are also scholars, that notions of individual property are the cause for much of the loss of indigenous community land and ecological resources, and that groups, communities and peoples need to be recognised as owning the land and knowledge of the land in common. On the other hand, many European and Asian FoE representatives in Malaysia argued for the need to include, and protect, 'individual' human rights, because in their experience, these are so often abused. By including both individual and group rights in the Vision and Mission, the activists do not make it clear how both can be nurtured simultaneously. In the subsequent two BGM and inter-BGMs that I attended this dilemma was not addressed specifically. But the importance of both was maintained in 'marketplace' discussions, especially by the FoE Colombia representative. Activists project the hope for something that does not yet quite exist (at least not in the knowledge or imagination of the activists) into the notions of equity, network, federation, being a grassroots federation, participation in decision-making. Flatness is one experimental gauge for these notions.

However, the flatness of the federation is probably exaggerated. The notions and rhetoric of flatness are devices that sometimes hide hierarchies. Seniority, personal charisma and patronage are obvious ways in which different FoE groups and different FoE activists have more weight than others in their decision-making discussions. As far as I could tell, some activists in FoE discussions are listened to more willingly and given more support than others. If powerful persons at international meetings give attention to specific people, then they gain more influence as a result. Activists with more power were sometimes those with more experience or seniority within FoEI. At other times, they were those who were more eloquent, more outspoken or who could more easily make themselves understood by many people in the room. However, for each decision different factors were at play. At times eloquence, but not seniority, held sway. At other times, even if the speakers were not particularly eloquent, were often interrupted, or found it hard to speak either in English or in Spanish, factors such as long-standing involvement in the federation lent support to their position as against the positions articulated by other speakers.

A simple story from my fieldwork illustrates how newcomers to the international meetings may have little influence, while long-standing and well-liked members of the FoEE office have more influence and may be listened to more. It was the first time that Janisa, from Latvia, has participated in a FoEE AGM. It was the last day of the SVPP workshop before the AGM proper began and the discussion at hand was about finalising the wording of the ten-year strategic plan for FoEI that the European groups would present to the SPTT team to collate with those of all the other regions. The statement being discussed referred to a suggested FoEI strategy to address the different, sometimes conflicting interests of imagined 'local' communities and an imagined 'globe'. The statement read 'global vs. local environmental interests'. The chair of the session, asked the plenary whether we agreed with the 'vs.'; he was not happy with it. The 'global', he said, should not be portrayed as being in conflict with the 'local'. Janisa, sitting next to me, raised her hand, was given the floor by the chair, and suggested 'the global *in relation to* the local'. No one picked up on this suggestion and the discussion went on. A few minutes later, Thomas, a long-standing member of the FoEE office and famous for his marriage to Dagmara, also a respected member of the FoEE office, raised his hand. He suggested the same change that Janisa proposed: 'the global *in relation to* the local'. Most of the people in the room agreed excitedly to the suggestion and it was adopted. There were no particularly bad or good relations between Janisa and the rest of the group. The only difference was that the well-loved activist was listened to more than the newcomer in this case.

At the international meetings, those with seniority and experience of FoEI are what one of the activists described to me as 'the silverbacks'. This activist happened to be an Austrian anthropologist. The silverbacks are those who dominate discussions at plenary meetings. Anthony explained to me that he does not feel comfortable speaking in plenary meetings. By default, this means that those who do not speak cannot contribute to the same degree to the outcome of the plenary discussions; as many FoE activists are aware, silence is assumed to be agreement. Anthony said that the smaller break-out groups were introduced through the SVPP in order for more people to have the space to contribute. Up until the BGM in Switzerland in 2002, the international meetings only consisted in plenary meetings. However, the influence of the 'silverbacks' is great at international meetings even now, and these discussions are anything but flat.

Kaarina's story complicates the matter. As I described in Chapter 6, Kaarina was a first-timer at the BGM in Abuja in 2006. There, she asked people not to take important discussions to the 'corridors'. Not only was her point agreed to during the meeting; it was also felt that her rebuke made an impression on many people. Her comment reminded people of the principles of equity and inclusion enshrined in FoEI's 'Mission and Vision', and which at that meeting were hanging on handwritten cloths in the plenary room. Kaarina's point was not just heard; it was supported. By appealing directly to FoEI ideology and 'the process', members present were reminded of the mismatch between their previous actions and their stated ideals.

The notion of network has been criticised because it flattens relationships that may not be flat and, as such, obscures power imbalances. However, in FoEI, the notion is actively used—precisely because of its connotations of flatness (and therefore equity)—to remind activists of their ideals, in the way that Kaarina did. It is a goal to work toward. Nonetheless, even though activists hope to develop flat, equitable structures, hierarchies develop. Though initially obscured, in many cases it is only a matter of time before those hierarchies are noted specifically by activists, who point to the discrepancy between their ideals and what is done in practice. The causes of the 2002 crisis show how the flattening power of the notion of a network or federation enabled those groups who perceived the lack of flatness in FoEI to have a radical effect on the course of events that followed. The concomitant ideals of a federation, of a network and of transparency—being against corridor lobbying—is what gave Kaarina, a first-timer, clout, or followship (Legat 2007), in the plenary discussion.[3]

Decentralisation

A network form, although flattened, can still depict relations of power through the density of nodal connections. The more connections a single node has the more relationships can be mobilised in support of that person (Boissevain 1974, 1977, 1979; Eriksen 1998). Accordingly, if one or more nodes have more connections than most other nodes within a limited set of relationships, then those nodes are more powerful; they have access to more resources. Imagine a diagram illustrating specifically the density of relationships. The nodes with more relationships will be pictured with an aurora of connections surrounding them, thus:

In this diagram, access to resources, which in this case lie in the number of relationships that can potentially be mobilised, is *centralised*. Notions of powerful centres and subaltern peripheries are also common in colonial and post-colonial discourses. FoE activists, especially those who are also scholars, are well versed in the critique of centralised decision-making structures. A network may flatten hierarchies, but to avoid a concentration of power a network must also be decentralised, activists argue. Therefore, the ideal of a flat structure is coupled in FoEI with the ideal of decentralisation, as well as inclusion in decision-making processes.

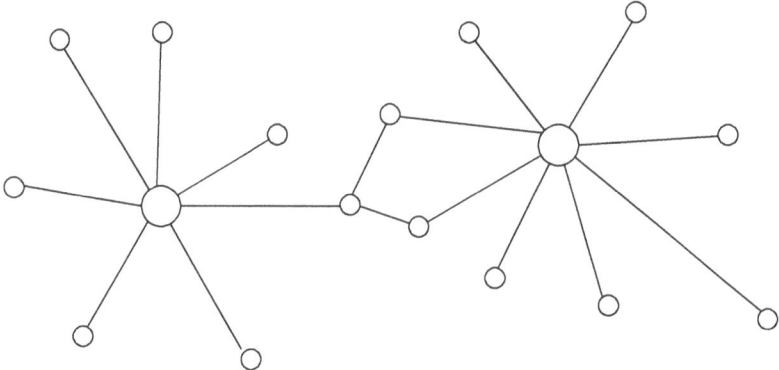

Figure 9.5 Nodes with more relations.

Probably more than 'flatness', decentralisation is a recurrent theme in the discourse of FoE activists vis-à-vis what is expected from the structure of FoEI. As I have mentioned before, a common comparison is with Greenpeace, which is considered centralised and 'top down', whereas FoEI is considered decentralised and flat. The critique in 2002, that the FoEI Secretariat was taking decisions without incorporating the different regions' positions, led to the SVPP. The different discussion and decision-making process of the SVPP is aimed at spreading—decentralising—decision-making power in the federation. The example I gave in Chapter 4, of Roberta, who was developing the idea of having Secretariat facilitators based in different countries, points to continued imaginings and experiments for further decentralisation.

Part of FoEI's mission is to build a society free from neo-liberalism and corporate globalisation. It is ironic, therefore, that decentralisation is also aggressively supported by promoters of global capitalism (multinational and transnational corporation [TNC], the WTO, the World Bank, the IMF, regional banks). Since the Second World War, there has been a major shift in the political environment from protectionism to reducing barriers to trade and capital flow. Specifically the WTO, and in its previous form the GATT (General Agreement on Trade and Tariffs), has facilitated the phenomenal growth of transnational corporations (Ietto-Gillies 2005) which are by definition decentralised, outside of the jurisprudence of any single nation-state (Druker 1997).[4] FoEI, however, led a big campaign against the Central American free trade zone. Free trade zones allow trade to take place without constraints between the countries in the zone. Without a free trade zone agreement, the central governments in a region control import and export levies. Similarly, liberal and neo-liberal policies calling for reduction in the national centralisation of services by privatising them are said to be reducing the centralisation of national economies.

The discussion at the FoEE AGM in Tartu about the relationship between 'the local' and 'the global' addressed this very issue. Were 'local communities' (as used by the activists) the only form of government, it would be impossible to negotiate about the 'global' environmental threats that FoEI is also concerned with, such as acid rain (in the 1980s), whaling and nuclear testing (also in the 1980s), and more recently climate change and various forms of inhumane working conditions associated with multinational brands. However, the value of decentralisation in the federation is not commonly questioned. Furthermore, beyond the questionable applications of decentralisation, aspects of centralisation can also be found in the Federation.

Here are three instances where the influence of the ExCom and the Secretariat can be understood as 'centralised' because its members have control over FoEI-wide contexts.

1   Imee, then the members' coordinator for the IS, had the decision-making power to refuse a table to FoEE at the Swaziland marketplace (although FoEE is listed as a 'group' on the FoEI website).[5]

2 The documents from regional SVPP workshops, such as the one I described earlier, are primarily 'collated' by the IS. Many activists who had worked on the regional SVPP strategy for FoEI in Estonia complained that the strategy maps presented in Abuja did not seem to include any of the points from the European regional meeting. A number of European groups, including FoE Malta, asked the director of the Secretariat to circulate similar documents prepared by the other 'FoEI' regions, but so far as I know those regional strategies were received neither by FoE Malta nor by FoE Brazil.

3 The ExCom meetings are not open to FoE activists, even as silent observers.

Decisions that are collectively made at the international meetings, on the other hand, are checked for decisions taken unilaterally. After the 'marketplace' discussions in Abuja, where I argued for the rights of non-humans to be included in the 'bubble' of democracy, I found that my point had not been included in the final flip chart hung up in the plenary. I had misunderstood 'inclusive discussions' to mean that all points were to be included, rather than only those on which there was consensus. So I simply added my point in felt pen to the flip chart. When the flip chart was presented to the rest of the plenary, my added note was immediately identified by a number of others who had been in that group as not having been agreed upon. I had to own up to having added it afterwards. Though I argued my point again, it was not supported and I was rebuked for not following 'due process'.

When such unilaterally taken decisions cannot be traced, it is harder for this checking to occur. It is also important to temper this critique with the frustration that the members of the Secretariat face daily. If activists from the member groups took a more active role in the work undertaken by the Secretariat, then there would be less need for the members of the Secretariat to do so much work themselves. Even this is fraught. Imee said that if only a handful of activists from a limited number of groups made an effort to 'check' the work of the Secretariat this would be seen as their having more power in the Federation. This seems to be an impossible situation. Access by member groups to the Secretariat and the ExCom, and participation in their work, implies having more influence in the federation. In order for balance to be restored, then every FoE group, or at least enough groups from each region, would need to take the same amount of interest (read as work, time and effort) in the Secretariat. This simply does not happen. For many years, the main effort of the Secretariat was directed at eliciting interest from the FoE groups in FoEI and in the work of the Secretariat (Link Archives, Amsterdam). Experienced members of the Secretariat argue that their actions, which look like 'control' and 'centralisation' in the description above, are in the best interests of the federation. A number of activists who are not in the Secretariat worry that without the work of the Secretariat and the ExCom, the International Federation would soon split apart. The ideals of decentralisation and flatness are sometimes de-emphasised, and instances of centralisation are forgiven, as a compromise to keep the federation working.

Inclusion is almost the other side of the coin of decentralisation, according to FoEI's Mission. Due to space limitations, I cannot develop an analysis of inclusion in much detail. However, it is important to make a few observations here. In order for decentralisation to be achieved, ideally all the FoE groups' positions should be included in decision-making. The rhetoric of inclusion in FoEI is expressed 1) in terms of 'ecological and cultural diversity', 2) in the emphasis placed on groups being established before they join the federation, and 3) in groups being encouraged to maintain their specific knowledge and approach to environmentalism. Homogenisation, once groups join the federation, is not encouraged and most groups do not change the name of their organisation to Friends of the Earth (see www.foei.org). However, FoEI is not universally inclusive. FoEI does not include corporations, employees of corporations, or industrialists. When a member of FoE Nigeria left to work with Shell, she was immediately excluded. In both FoE Malta and FoE Brazil, activists keep an updated list of companies from which they will neither seek nor accept funds. In FoE Malta, for many years, any support from the government was also avoided, because of the obligations of patronage that are tied to such support. With every decision to work with someone, with a group or movement (such as Via Campesina), or locally with Churches, and not with someone else (like the move away from UN fora), an exclusion is affected. I make this obvious point only because FoEI aims to 'nurture ecological and cultural diversity' and 'to ensure public participation in decision-making'. This highlights the challenge, and the lack of clarity, over what 'inclusiveness' of the sort described in FoEI's Mission could be in practice. Including different systems of government in FoE strategic thinking has proved difficult. Divisions about different economic ideologies, and about capitalism in particular, are even stronger and the notions commonly associated with it, such as entrepreneurship, are often held in suspicion if not rejected outright. Finally, the external expectation and pressure on FoEI to uphold coherence creates even more challenges in their exploration of the notion of 'inclusion'.

Consensus

Achieving consensus as opposed to reaching a decision through a majority of votes is an ideal that became formalised the SVPP. All the SVPP documents produced so far—the Mission, Vision and Values, the strategic map and sub-level maps, and the initial implementation plans for the next couple of years—have been approved through consensus decision-making processes such as the inner/outer circle method. The decisions during the 'business meetings' of the GMs are, however, taken by voting, although 'extended discussion' is part of the process for 'sensitive' decisions, such as the retraction of Amazonia Brasileira's affiliation. If in SVPP discussions a consensus is reached, then it is imagined that inclusion has been achieved in practice. In theory, consensus is evidence of inclusion. In practice, the ideological elevation of consensual decisions, and the insistence of the SVPP coordination team on consensus, sometimes have effects that are the opposite of the desired inclusion, flattening and decentralisation of power.

In discussions in GM business meetings, compromises between opposing positions are sought before a vote is taken. Amigos da Terra Amazonia Brasileira which, until 2006, was an affiliate member of FoEI, working in the Northern Amazon in Brazil. A motion at the Abuja BGM to 'immediately retract Amazonia Brasileira's affiliation' was modified because of discussion before a vote was taken. The motion was changed to a conditional statement that 'if Amazonia Brasileira does not cooperate with FoEI to address the alleged problem of confusion, *then* its affiliation would be retracted'. With this amendment, which included a compromise on the timing and condition of the retraction, the motion was passed.

Compromise in consensus decision-making in the SVPP process, which I participated in, did not appear to be so different from these pre-voting discussions. In fact, to confirm that a decision was actually reached, the facilitator or chairperson of any discussion would ask for a show of support or otherwise. The difference between the closure achieved in 'consensus' and in voting lies in how such support is counted. 'Votes' are counted, whereas in consensus decision-making 'support' is estimated on the basis of an informal show of hands, where the hands are not counted, or through positive or negative sounds of clapping or conversely booing. In these informal voting procedures, as in discussions, silence is automatically taken as approval. Moreover, unlike in voting, if even one person does not approve a proposal and speaks up against it with sufficient eloquence, then the discussion is reopened.

However, it is not assured that objections will always be voiced. Many activists do not feel comfortable speaking in plenary meetings, either due to language difficulties, or because of what they consider to be lack of experience, which is why the breakout groups were introduced in the first place. Moreover, as evidenced by the example of 'local vs. global' described earlier, when activists do speak they may not be heard, depending on their ascribed position in the meeting. For instance, in the debate about democracy in Abuja, Demian made a number of contributions, but his comments were not taken as seriously as Rainer's. Finally, some informal ways of determining support for or dissent from a proposal make it very difficult to express disagreement.

At the FoEE AGM in Estonia, the plenary had to vote on the overall strategy for FoEI proposed by the European region that we had been working on for the past three days. The method of determining agreement in the spirit of consensus, after discussion of the strategy, was as follows: with all the chairs removed and everyone asked to stand in the middle of the room, the facilitator asked all those who agreed with the overall map to move to the right and all those who disagreed with it to move to the left. I disagreed with some aspects and believed there was need for more discussion (specifically on the sedimentation of FoEI into regions, which I believed would increase disagreements among groups based on regional factions, whereas many groups across different regions shared many of the same concerns). So I decided to move to the left. As it turned out, although I knew there were other people who wanted to carry on the discussion, I was the only one left standing on my own. Such processes make it difficult for people to speak up.

However, when an individual objects in such a consensus decision-making process, this may effectively become a veto. At the Abuja 2006 BGM the debate over inclusion of the word 'democracy' in the overall map was the most obvious situation in which the power of veto in consensus decision-making was at the forefront of activists' concerns. The debate was long and heated (also see Doherty 2006). When a number of European groups, the most vocal of all being the representative of FoE Germany sitting in the outer circle, called for 'existing democracies' to be added to the overall map, the FoE Colombia representative, sitting in the inner circle, objected. The debate thereafter turned on the issue that the South American groups, including the Colombian group, saw the notion of existing democracies in her region as a veil for corruption and all the abuse they experienced in their countries. On the other hand, for many of the European groups, most of their work was directed towards governing structures that they, the activists, call democratic; structures that they consider do work, even if they have their problems. Some European FoE activists argued that if 'existing democracies' are not mentioned on the overall map (as they were not mentioned in the Mission, Vision and Values), then their work would not be represented in the official documents of FoEI. The German activist, Rainer, said, 'We cannot find ourselves reflected in the map'. As I mentioned previously, the overall map was intended to be exactly that, a map of the current work of the different FoE groups in FoEI. According to Vera, the map was meant to be 'inclusive' of the different types of environmentalist work that FoEI encompassed.

When the debate became very heated, Rainer appealed to the chairperson of FoEI, Meena Raman (her real name), 'I tried to talk with my bad English, and I don't like the process, because I explained to you how I feel [about not being reflected in the map without democracy mentioned], and we achieved a compromise but one group keeps pushing and pushing against the wall'. Agata, the Colombian activist, replied, 'This is political not personal, I am not persecuting, I am pushing my region's political position, if we cannot see that, this process cannot continue'. Vera tried to facilitate: 'The word "compromise" is difficult—we should look for an inclusive strategy, so that we don't have to do everything, but so that there is also space for everyone'.

The discussion had been going on for five hours. In the end the clause on the overall map was approved as '*Create and build new democratic structures. Influence existing ones*'. A regional meeting was called the following day because the argument over 'democracy' created dissatisfaction with the process. Demian commented during the debate, 'Yesterday we spoke about inclusiveness, and solidarity. I don't feel it today.'

On the one hand, the argument for consensus decision-making, specifically for something like an inclusive map of the work within the federation, is compelling. How could anyone vote against a map describing the work done in the federation? To do so would directly contradict the other principles of the federation that prevent it from interfering with the work of member groups. The difficulty lies in situations when activists perceive a contradiction in what is being discussed. It was felt that the South and Central American groups' arguing against having 'existing democratic structures' in the overall map might lead to the same sort of situation that resulted in Ecuador leaving FoEI. That is, if they as South American groups did not support their so-called democratic governments because of corruption and violence, they would then be accused of not being as 'reasonable' as their 'European' counterparts who *do* endorse the system of their 'democratic' governments. For Agata (from FoE Colombia), if the FoEI overall map were to include existing democratic structures, this would contradict their work of unmasking just how *un*democratic the structures they work within are. For Rainer, if the overall map did not include existing democratic structures, then FoE Germany would be excluded from FoEI. What this discussion highlights is that for consensus decision-making to work there has to be understanding of the differences among the groups.

There is a stark difference between consensus decision-making and the consensus-building processes that happen in the smaller groups; the former carries a much higher risk of vetoes. In Chapter 4 I described how, during the smaller breakout groups, the different ideas proposed by participants are all written up on the working flip chart paper. When there is an apparent contradiction, efforts are made to bring a resolution to that contradiction by adding ideas or possibilities, finding ways to turn conflicting ideas into complementary ones. In all the breakout groups I have participated in, I have not once experienced a situation in which a contradiction could not be resolved. The difference is that in the process described earlier, the discussion was about a document that had already been compiled, and this is where feelings of 'not owning the process' may arise. The discussions about the strategic maps were more akin to voting on ready-made motions than to building consensus, as in the breakout groups. Of course, the argument for processes like the inner/outer circle is that it is difficult for the large number of people present in the plenary to compile the different ideas developed in the smaller groups. However, the solution to the 'existing democratic structures' debate was that a group of activists from different regions moved aside and devised a proposal from scratch for that 'bubble'—in effect creating a new proposal. Even when motions are being discussed in the business meetings, counter-motions may be passed during the discussion itself, confirming that the process of idea-generation *during* the decision-making discussion is very helpful in creating agreement. In fact, there is at least one commercial organisation in the UK (Centre for Creative Management) that employs the same processes of consensus building used in the small breakout groups for very large numbers. Ironically, CCM have adapted consensus-building tools to help large for-profit corporations 'gel' when there is a merger between different companies.

Facilitation

Both the coordinators of Moviment Ghall-Ambjent/FoE Malta and of NAT Brasil/FoE Brazil consider their role to be one of leadership through facilitation. Their roles and the organisations they coordinate are very different. FoE Malta is primarily volunteer-based with a small annual turnover in relation to other ENGOs in Malta. Anthony, the coordinator of FoE Malta, is a volunteer: he has another full-time job and is completing a correspondence degree in environmental management. FoE Brazil has more full-time employees than volunteers, and receives enough funds for salaries, publishing, events, office space and overheads. Nonetheless, their respective notions of leadership are strikingly similar. Both coordinators have explained that their roles are exactly that: to 'coordinate' the different contributions of the activists (volunteers and employees) in their groups. They see their role to be the creation of space for other people to participate in the group. In practice, there are a number of instances when the role of coordinator becomes one of leadership. There are different responses to this in the two groups.

In FoE Brazil, there were two conflicting responses, each relating to a specific negotiation of interests. On the one hand, Andrea's approach to leadership was criticised by the previous (informal) coordinator. She was not giving the group 'strong leadership', the criticism ran, and as a result the preparation stage of the project for the new NAT office seemed to have taken much longer and to have caused frustrations between activists. On the other hand, when the requests of three activists to attend FoEI international meetings were repeatedly turned down, because 'their English isn't good enough', Andrea was also criticised. The three activists gave me their interpretation of the refusal: Andrea does not agree with their political positions and thinks that they will embarrass FoE Brazil with their views if they go to a FoEI meeting. They think Andrea is promoting her personal interests rather than acting as a facilitator.

In FoE Malta, there has not been the same open criticism. However Elisa, who was employed through the capacity-building project as a 'Change Agent', was always overworked and complained that she had to run after the other activists. Apart from the fact that the activists are volunteers, including full-time housewives and students, and therefore that all have other jobs, it could be argued that so-called 'strong leadership' could inspire more dedication.

Interestingly, Anthony's and Andrea's notion of what is involved in being a coordinator is very similar to the way the IS talk of their role in the federation as being one of facilitation. It becomes problematic with the IS because their facilitation role carries clear aspects and effects of leadership, which are not always made explicit. For instance, the IS proposes the processes for the SVPP. These then have to be discussed and approved by the ExCom. But in practice, since the ExCom only meets four times a year, most of the proposals for processes come from the Secretariat. At the international meetings, these proposals by the IS are accepted or rejected, depending on who is proposing them. For instance the process for the workshop on 'building a strong federation' at the Swaziland meeting, proposed by Roger, who is associated with FoEE and FoE EWNI, was rejected. On the other hand, the various processes such as the timeline, the Mission, Vision and Values, the inner/outer circle, the strategic map and so on are proposed to the plenaries by Vera, who is introduced as Indian, and are all accepted—even though the tasks she suggests are used in big corporations when there are takeovers or when organisations need to 'rebrand' (pers comm. Tamsin Valentino Client Services Manager, and with John Varney Uniform UK, Centre for Creative Management, UK).

Another instance happened at a FoEI workshop in Swaziland in 2007. During the round of introductions, the Director of the Secretariat chose not to introduce herself and her expectations of the meeting like everyone else. She was only there as a facilitator, she protested. This is an example of how not recognising the aspects of 'leadership' in facilitation leans towards lack of transparency. Due to limitations of space, I cannot go into this aspect in detail. However, I would like to make a possible proposal.

Resulting from my limited experience with consensus building in an applied anthropology project I propose that the position of the facilitator (his or her interests, wishes, desires, affiliations) should be stated, and that the facilitator should also be given a decision-making role. The facilitator, rather than only assisting the discussion, is granted an equal stake in the discussion (Gatt 2005). When, as happens in FoEI with national membergroup 'facilitators' and the IS, the facilitator is understood as the channel for discussion, but is not acknowledged to be a subject with interests and positions of his or her own, then it is more likely that these interests will be pursued anyway, and in a way this is inevitable, but in a veiled manner. Giving the facilitator the right to have a position, and the right to defend his or her interests, reduces the extent to which these interests need to be veiled. However, this also increases the responsibility of both the facilitator and other participants in the discussion to ensure that the facilitator is simultaneously managing to 'facilitate' the discussion process (see Gatt 2005 for details). I derive this proposal from the literature on reflexivity in anthropology, where the writer's assumptions are stated up front in order to allow readers to judge his or her bias. David Mosse (2005) has proposed a similar strategy for development projects.

Positions and Persons

What I have analysed as contradictions seem to arise from tension between the personal and the political. The tension lies between thinking of the activists as abstract political actors, playing their parts within carefully choreographed set-pieces (inner/outer circle, flattened networks, etc.), and on the other hand, as flesh-and-blood human beings subject to all the usual feelings of pride, embarrassment, hope, interest and so on. Behind the contradictions that I examined, there is a more fundamental one: that while 'playing the game' at these meetings means keeping the personal apart from the political, the merging of the two—making the political personal and vice versa—is at the core of what it means to be an activist. In fact, the need to separate the personal from the political is not clear-cut even at the meetings. We saw at the beginning of the chapter how Henry is working with the idea that to make 'inclusion' practicable the activists need to know and understand each other's personal experiences. Doherty (2006: 14) also notes how the very carefully planned discussion processes—inner/outer circle, breakout groups, market places, and other Open Space Technology tools—depend on the activists' contributing their personal experiences and thoughts, as opposed to the formalised position of the member group they represent.

The tension, then, can be understood to be produced by certain conceptions of 'an office' or 'a position' as necessarily separate from the interests of the specific persons who fulfil the role; a conceptualisation fundamental for bureaucratic praxis (Weber 2002 [1905]). What, at least in part, follows from this is the notion that bureaucratic relations are abstract and emptied of personal interests and relationships (Strathern 1996). In her study of NGO workers in Fiji, Riles (2000: 65) argues that personal relations are defined as the interactions that happen outside pathways of communication formalised in the 'network'. Therefore, the interactions that happen within 'the network' are understood as non-personal, as political. However, in FoEI the activists refute this distinction between personal and 'network' relations. They understand their entire lives to be part of their FoE work and the basis for solidarity developed during international meetings consist in their getting to know each other as persons (not representatives of this or that member group). We even saw how the daily work through ICTs is understood to be enhanced if the activists develop multidimensional relationships—visiting each other's homes, going on holidays together and so on. Thus, the tension is caused by apparently conflicting attitudes to how the activists should engage with each other.

Considering the multiple and simultaneous scales of emplacement that activists inhabit, as well as the environmentalist notion that everything is ultimately interdependent, the question of personal interests versus political ones is open-ended. It is a question the activists are themselves probing. What does it mean to have a 'federation'? What does it mean to take decisions about the federation in a situation where differences and personal experience are valued? What does it entail to search for liberation from all oppression (including that of non-humans)? I suggest that this tension is productive, and that it arises from the character of experimentation.

The SVPP has refined the activists' attentiveness to 'process' and 'politics' as part of their hope for an equitable and sustainable future. Here the mismatch seems to occur when neither the ideology nor the actions consistent with it are actually known. The ideal relationship between activists is couched both in personal and in abstract terms, because aspects of both are valued. However, activists do not yet seem to able to embrace notions and practice these valued aspects in principle and practice, without simultaneously creating tension and apparent contradiction. And yet the situations in which conflicts, tensions and apparent mismatches arise are not discarded by the activists. Personalism, at the meetings and in interactions among activists, is valued concurrently with ideas and structures of disinterested decision-making. Tensions and conflicts, from the perspective of coherence (such as Quakers' deference to unity, Plüs 1995), can be understood as failure. However, the defining quality of experimentation is also 'the freedom to fail' (Harkness 2009: 159). For Harkness, the "implications of experiment and play . . . are that designs sometimes 'don't work'" (*ibid*). Experimentation is a leap into the unknown and the unknowable (Harvey, cited in Harkness 2009: 160) and as such, failures and successes can only be known in retrospect, after the leap.

What environmental 'justice', 'inclusiveness' and 'equity' actually could be is partly unknown to FoE activists. They consider that they live in situations where these do not exist, or do not exist perfectly. FoE activists are in the process of experimenting *for* 'environmental justice' through ideas and experiences of 'decentralisation', 'inclusion', 'flatness' and so on. These notions are therefore also experimental aims. The words are tools in the sense that they help them experiment with ideas and practices as they go along, fine-tuning them on the basis of experience and either 'reclaiming words' (Demian, from FoE Hungary) such as 'democracy' or exploring new possibilities as yet unnamed. It is worth quoting Harkness at some length here:

> [the] argument here is that the creative act and the consideration of possible futures require both that people should try to 'know the implications of their contemplated acts' (McHarg, 1997: 55) and that they should engage in openended play and experiment. Earthshippers are not blindly experimenting with things they do not understand; they are participating in ongoing creative explorations of their craft in their environments. Their craft is one that is oriented to incorporate the results of such explorations and to know and consider the places and people that it affects.
>
> Harkness (2009: 159)

The *craft* of FoEI activists, in this case, lies in imagining notions of possible futures. Terms such as flatness, facilitation, inclusion, decentralisation and consensus are vehicles for part of this experimentation. That they only fit uncomfortably for now is the grist that allows the activists to evaluate the experiment as they go along. Richards (1993) notes that, as well as allowing for mistakes, an important part of experimentation is careful observation of what is happening around us. In the activists' experiments there is a back and forth movement between experimenting with practices and processes of decision-making which feeds into experimenting with concepts.

Discourse may, whether strategically or not, be part of this hopeful, experimental exploration. It is hopeful in the sense that by using certain words and notions, such as flatness, decentralisation, inclusion and facilitation, FoE activists are trying to bring about a change towards a better way of formulating policies and taking decisions. The 'better way', however, is not easily found; there are no ready-made processes for the sorts of inclusion and celebration of diversity that FoE activists are searching for.

Schurman and Munro (2006) have argued that a vital part of the environmentalist movement is specifically this experimentation with new notions, or as they call it 'thinking work'. It is important to note here how, rather than understanding 'thinking' as a separate sphere from practice, the notion of 'thinking work', like the Earthship-builders' work described by Harkness, is part of the activists' ongoing work within and response to their environment. Thinking work, like imagination, is a practice that has 'a quality of attention that is embodied in the activity itself' (Ingold 2000a: 417). This can be seen when the activists' observations of decision-making experiments and the imagining of possible futures feed back and forth between each other for further experiments.

To conclude I suggest that the activists' experiments contribute to Talal Asad's (1979, 1983) critiques of classical symbolic anthropology. He argues that human ways of life, people's perceptions and actions, cannot be understood as determined by a system of meanings. If social structure is understood, as it classically was by Leach, Douglas and Geertz, among others, to be an integrated totality of social classifications and meanings, and if the origin of concepts is social (that is, if cognition is determined by society), then it becomes impossible to specify how social change can occur (Asad 1979: 611). I would like to add that if perception and action are mediated by cognised system of meanings, then there can be no logical space for mistakes, let alone for experimentation.

It may seem anachronistic to refer to a critique of symbolic anthropology dating from the late 1970s and early 1980s. The 'practice turn' (Ortner 1984) in anthropology seems to have taken on board the sorts of critiques made by Asad (1979, 1983) and Bourdieu (1990), among others, with the result that entirely different sorts of questions are now being taken forward. Nonetheless, key texts on environmentalism, such as Argyrou's (2005) *the logic of environmentalism*, firmly uphold the assumptions of symbolic anthropology.

For instance, Argyrou finds the idea that people attune to different aspects of the environment to be problematic because the perception of that environment, for him, is '*always already* a cognitive reconstruction of one sort or another' (2005: 95 emphasis in the original). Where, I wonder, would thought experiments fit into Argyrou's understanding that perception is 'always already' a cognitive reconstruction?

Notes

1 This discussion process is part of what is called Open Space Technology (OST).
2 Taken from www.foei.org, accessed on the 20th February 2008. This is repeated from Chapter 1 for convenience.
3 A comparison with band organisation could shed light on the differences between informal, and possibly shifting, hierarchies as opposed to formalised ones (see Legat 2007; Bird-David 1999, 1994; Ingold, Riches and Woodburn 1988). However, due to limitations of space, I cannot explore this aspect in more detail here.
4 In fact, the legal issues of jurisprudence, of which a country's laws are to apply when an office has headquarters in one or more other countries, are at cutting-edge legal research (Borg Barthet 2010).
5 www.foei.org/en/who-we-are/member-directory/groups-by-region Accessed 6th August 2010

10 Epilogue
Keeping the Conversation Open

The chapters of this book have followed the typical progression of a FoE activist, from joining as a newcomer to a national member group (Chapter 4), becoming incorporated into that group (Chapter 5), to going to the international meetings and being a newcomer to FoE International (Chapter 6), to returning to his or her respective home country and maintaining relationships with other FoE activists through ICTs (Chapter 7), to encountering different aspects of the work involved in being an established activist, which includes dealing with different sorts of entities (Chapter 8) and finally to being in a position to critique the Federation (Chapter 9). In Chapters 2 and 3, the questions and approaches that informed my research are akin in nature, even if not in content, to the motivations that drive people to join a FoE group in the first place. People develop dissatisfaction towards what they see in the world around them and formulate the beginnings of a way to improve these situations. With that embryonic formulation, it becomes clearer which groups of people they might have some sort of affinity with, even if this may be fuzzy at the beginning. This is how people come to think of joining a FoE group rather than some other organisation. Of course, these different formulations are part of the uniqueness that each activist brings with them to their newly joined FoE group, and that is valued because it is understood to bring a diversity of skills and ideas to the respective groups. Chapters 2 and 3 are not the formulations I had before beginning my work with FoE Malta which, as I have stated, began in February 2003. These two chapters are the product of the whole process of my engagement with FoE groups. They have been constantly revisited, right up until the final stages of writing the book. They do, however, include some of the ideas that I have carried with me for many years, even before joining FoE Malta. They are placed at the beginning of the book because the approach proposed therein underlies the observations and the type of analysis that pervade the rest of the chapters.

Chapter 2 proposed ecological phenomenology as an ontology based on the work of three scholars, Tim Ingold, Bruno Latour and Donna Haraway. There are two key points to this phenomenology. First, the interactions between humans, and between humans and non-humans, in other words between all the constituents in any milieu, are *generative*. These interactions, in which persons and things engage, are what actually constitute them, and thus constantly bring the world into being. Human perception, in this understanding, is also part of an active, generative process; perception therefore is not the cognitive domain of human reception and interpretation of inchoate stimuli from a disparate, natural, domain. Second, the world is a single arena that is continually in the process of becoming, there are no unbridgeable gaps between mind and world or between society and nature. The framework of force fields and vectors offers a simple means to conceptualise the effect of any given situation. This effect can be understood, and traced, by mapping the different types of force and the ways these interact with others, in other words through the directionality implicit in vectors. For example, in Izabela's initial experience, the EU was a goliath constituted by the various vectors of symbolic power. Names and ideas, together with the practical power of doormen, secretaries and other officials, as well as the architecture of the buildings and the particular colours and materials used for such constructions, worked together to create an impenetrable entity that was very real in the effect it had in limiting Izabela's possibilities of action. In accordance with the generative function of relationships in ecological phenomenology, the force fields I have presented in this book map the constitutive forces of some significant vectors.

Following the way a newcomer to a FoE national member group is first introduced to the organisation through history texts, in Chapter 4 we took a first look at the constitution of FoE groups through these texts. They create coherence for FoE Malta, FoE Brazil and FoEI through two distinct understandings of time and history. The first understanding of the history of the organisation, concerned with founding dates, chronology and lists of activities, is part of a vision that depicts organisations (or nations) as solid, discrete entities moving through empty time. With the second understanding, however, the activists experience their participation in the organisation as growth, not in the sense that the organisation gets bigger, though this is sometimes the case, but in the sense that it lives and changes in the way an organism grows. In the understanding of growth, there is duration, and rhythms that change according to what and how things are happening. Though there may be moments of heightened activity or pressure, even quiet, daily and often repetitive goings-on constitute the organisation and its history. In the growth history of organisations, founding dates and organisational boundaries are less emphasised. The former, chronological, understanding of the organisations' history is important to ensure continued support and funding from their environments of legitimacy. The latter, organic, understanding of the organisations' development is important because it embodies the sense of continuity experienced by activists. In fact, both notions are incorporated into their documents.

In Chapter 5, we saw how the FoE groups are understood to be constituted by their activists and how FoEI is the sum of all those activists. The lives, experience and knowledge of activists, even outside of and prior to their specifically FoE activities, are also incorporated into the constitution of their respective national member groups. Therefore, the life histories of the activists are central to understanding how FoEI hangs together. The activists have 'co-responding' notions of life as a path, a path that they consider themselves morally obliged to forge through their actions. The activists are disciplined in reflexively monitoring their actions to forge this life path. They have, in other words, a technology of the self (Foucault 1988), or a *disciplina* (Asad 1987). They guide their disciplined reflexivity, not according the strict and unchanging rules of a religious text (Van der Veer 1989) or the dictates of a leader (Asad 1987; Foucault 1988), but through an eclectic and flexible responsiveness to the new and unexpected circumstances they face in daily life.

Most activists recognise that they occupy relatively privileged positions. They compare their wealth with the people they have got to know through their work with FoEI, as well as with people and situations they knew before. In many instances, this led them to join FoE groups and not other, more conservationist ENGOs. The recognition of these lifestyle differences has led many of them to want to use their position as urban, educated and relatively wealthy people to improve the quality of life of others less well off than themselves.

The annual FoEI international meetings, which are the focus of Chapter 6, are settings during which, and through which, activists develop and reinvigorate their sense of belonging to FoEI. The collective sense of belonging is central to the imagined constitution of FoEI, along the lines of Anderson's (2006) notion of imagined communities. However, the meetings are also central in the actual constitution of FoEI since they are the primary settings in which activists get to know others from different national member groups in person. They may have heard of these other groups, or emailed with FoE representatives, but the knowledge they have of them is attributional, in the sense that they are known to each other as this or that national representative. The personal relationships developed during international meetings are essential because after the meetings, when the activists have returned to their respective countries, they go on to develop work for FoEI together, with very different outcomes than would have been possible, had they not got to know each other.

Activists come from very different backgrounds, and bring with them very different ways of perceiving environmentalism and the world. They speak many languages. However, the intense, bubble-like atmosphere at these meetings creates a common—or at least a 'co-responding'—ground of experience in which such personal relationships can develop in spite of so many differences. The atmosphere and consequent opportunity to develop multidimensional personal relationships, and thus a sense of belonging different from that of the imagined community, emerge from the collective rhythms they engage in.

In Chapter 7, we follow the activists back to their respective countries, where much of their joint work is carried out by means of communications technologies, primarily by telephone or email. Activists in the same room may often prefer to use email or Skype conference calls rather than walking over to the other person in the room, because this allows others, not in metric proximity, to participate. Through such communications technologies, the activists are present to each other just as much as they are in presence in metric proximity. Different types of presence are generated according to the different affordances of the medium within which interactions occur. In metric proximity, air is the medium that allows a variety of ways to respond and perceive the presence of others, including seeing each other, hearing each other, feeling pressure as well as actual touch and smell. Over the phone, even if one cannot see (although video conferencing is improved and available), smell, touch or move with others, the transductive properties of electric and fibre optic cables, and the technology developed to create such transduction, make it possible for us to hear and to see others in (at least to our senses) real-time. That is, we can perceive each other's reactions to us with the same speed as we would perceive their voices travelling through the air in metric proximity.

Such real, not virtual, presence—co-generated by the practices of activists, the technologies, and the affordances of the materials that make such technologies successful—are essential to the capacity of FoEI to hang together across the over seventy member countries. Even very recent attempts to address the question 'What is a medium?' assume a contrast between the immediacy of face-to-face communication and communication that is mediated by technology, images, the spirit world and so on (Eisenlohr 2011; Hirschkind 2011; Engelke 2011). By attending to what actually goes on when activists communicate at a distance, however, we have been able to dispense with the mystification of the 'virtual world'.

In Chapter 8, we saw that the entities that activists come up against in their FoE work are not only other people. At times activists feel that they are engaging a Leviathan, such as the EU, or the FoEI Secretariat, or other, apparently more powerful FoE groups. In other situations, the activists deploy 'FoEI' in such a way that it comes across as a solid entity, precisely because the organisation as a 'thing' is perceived as fragile. In these situations, transient and intangible relationships are only successful in creating a supra-personal entity, or a certain fixity, only because of the distance of the perceiver; distance that can be caused, in the cases I analyse, by a lack of familiarity.

In other situations, such as when FoE Malta participated in a capacity-building project, although FoE Malta activists were learning about their 'experienced twin', FoE EWNI, this did not make them more familiar with the actual daily workings of that other organisation. As a result, FoE EWNI, and its representative, remained all-powerful in making possibilities available to the FoE Malta activists. The latter had unprotected backs to the workings of FoE EWNI; that is, they did not direct their attention to the multiple relationships and diverse aspects that are included in FoE EWNI's constitution as an organisation. Rather, they had their attention directed towards the 'accomplishments of FoE EWNI'. As a result, they did not question this situation as much as they had been accustomed to questioning other situations. In Chapter 8, I trace some situations in which different directions of attention, together with other vectors, can explain how suprapersonal entities have actual effects on people as real things, and how those same entities change, depending on the direction of attention.

Finally, in Chapter 9, I compare the ideals that FoEI claims to stand for, and consider whether these ideals are achieved in actual practice. The ideals are flatness, decentralisation, diversity, inclusiveness and facilitation. In practice, there are hidden hierarchies, centralised forms of decision-making and exclusions, and facilitation often veils the attempts of particular individuals to push their own ideas, rather than simply moderating discussion among others. There also seems to be a contradiction between forms of decision-making that expect activists to renounce personal interests and other aspects of those same situations in which they are encouraged to engage as interested persons.

I argue that these contradictions arise because FoE activists are experimenting with forms of decision-making that incorporate valued aspects of both these stances: on the one hand, the lack of oppression implied by the notion that one should not pursue personal interests, and on the other, the value given to each person (as opposed to anonymous, non-participatory processes of decision-making). In these experiments, the notions of flatness, decentralisation and so on are imperfect criteria against which to judge activists' efforts. Ideas, here, are vectors, in that they have effects on the constitution of FoEI.

Answers, Implications, Challenges

One of the questions that drive this book concerns the multiple claims on the world that anthropologists and FoE activists, even in their Vision and Mission, talk of as 'cultural diversity'. How can we reconcile diverse human ways of life in a way that does not relegate such diversity to 'belief', and all the while recognise that the same environments afford different ways of life? The ontology of ecological phenomenology that I presented in Chapter 2 provides an answer to this issue. The world and our environments are constantly being made, and our acts of perception, as well as all our other practices, participate in that constitution—which at the same time constitutes our own being. Because of this, different practices create different affordances in those milieux. This is not, however, to say that people culturally construct their environments. These environments include non-human, non-cognitive aspects that have their own affordances and provide constraints and opportunities for the practices we engage in. Although different practices (with various human, non-human, cognitive and non-cognitive participants) co-produce these environments, the fact that the non-cognitive participants have their own ontological presence means that people can come to perceive the same affordances through the education of their attention (Ingold 1992).

This has implications for discussions among FoEI activists, as well as for other situations in which there is a need for cross-cultural negotiation. That different perceptions of the environment are not simply beliefs, but arise concretely from processes of mutual constitution, implies that the claims that 'other' views are only matters of 'belief' are not so easily accepted. In addition, that different perceptions of the environment can be learned, through the sorts of education of attention that anthropologists engage in during fieldwork, offers a practical tool for such negotiations. Of course, this happens in the form of exchanges in FoEI, as well as in many other situations. However, what ecological phenomenology adds is that those acts of perception are also creative. This has the further implication that what is focused on—that is, what affordances are given attention—can be deliberately directed at creating particular situations. Therefore, it would be possible to run negotiations by first creating commonly constituted situations. This, I argue, is what is happening at the international meetings of FoEI, described in Chapter 6.

The second question of the book concerns the different dimensions of power including symbolic power, material resources, personal power and the agency of non-humans. The framework of fields of forces and vectors allows us to take into account different qualities of power at the same time. This is because anything can be a vector, and it is not the quality of the thing that the vector represents but the effect that it has that is important. Of course, if the quality is what specifically creates the effect, then a vector that represents the effect will reflect the quality as well. In Chapter 7, we saw how the quality of electronic technologies that make possible the transduction of voice, images and data, only effects 'co-presence' if the people using those technologies attribute a certain value to communication. If, like the Amish, only presence in metric proximity is valued for community building, then the telephone would not be successful in creating acceptable presence. On the other hand, no matter how willing the FoEI activists were to have a Skype conference call, the narrow bandwidth being used by some of the participants prevented the success of that meeting. In different situations, symbolic value or power is a stronger vector than the affordances of material things, and vice versa. At other times it is how these, and other distinct forms of power, coincide, intersect, and counteract each other that produces effects. These can be pictured by using the notions of vectors and force fields.

Vectors and fields of forces, while presenting constitutive movements within a single, common, picture, nonetheless prevent the ontology of ecological phenomenology becoming a 'monism'; spiritual dimensions and disembodied minds (Porath 2007) are present in my ethnographic account. In the framework of fields of forces and vectors, these do not need to be reduced either to belief or to be explained as material, they do not lose their difference, because what is important is that their effect can be perceived— and by this I mean perceived by the people we work with in our studies. Where there are different perceptions of the same phenomena, these can be attributed to differences in the education of attention, which is what, in effect, the process of anthropological fieldwork involves. Admittedly, this does not explain how different perceptions emerge in the first place. These sorts of questions can only be answered in relation to specific situations or phenomena.

In ecological phenomenology, a key premise is that any phenomenon is constituted by those participating in its occurrence, who are themselves the momentary outcomes of similarly constitutive interactions. What, then, should one focus on in research? Force fields and vectors also provide a guide to addressing this question in practice. Depending on the types of vectors and the force fields that can be recognised by their effect, one can decide whether a phenomenon, a constituent or an actor is relevant or not to the questions at hand. For example, in relation to wanting to pay attention to the organism-person (Ingold 2000b), I was asked whether the guts of FoE activists are important to understanding how FoEI as a thing is constituted and maintained. Off the cuff, I would say 'no'. However, guided by the happenings in fieldwork, it was obvious that the activists' breathing, digestive, sleeping and dancing rhythms were essential to understanding how they developed a sense of belonging and commonality at FoEI international meetings. These commonly shared rhythms, which are both to do with the organism and the meaningfulness, for example, given to dancing at the meetings, are directly relevant to constitution and maintenance of FoEI as an organisation. It is open to debate whether this approach can inform a post-human anthropology (Ingold 2011b).

The third question of the book encapsulated the challenge of grasping the phenomenological aspects of globalisation without—and this is key—reducing supra-personal aspects of 'the global' to nonetheless local phenomena. For instance, the notion of re/deterritorialisation that Inda and Rosaldo (2002) develop, is based on the idea that notions, images and forms need to be *removed* from particular locales, even as they are simultaneously re-territorialised elsewhere. In FoEI, many practices are occurring in places that are not local, in the sense of being in metric proximity, nor are ideas or images launched forth into the world through mass media to be re-localised and re-interpreted somewhere else. Rather, most of the daily work done in FoEI occurs through communications technology across many different countries, metrically both distant, and close (for instance, in a Skype conference call where two people are in the same office together, but the other participants are spread around the world). In these situations, policies, ideas and plans are jointly created. Of course, there is necessarily a degree of difference in the interpretations that people express in speaking or in writing. However, in discussions by email or on the telephone, it is often those interpretations that are debated.

In both Chapter 3 and Chapter 7, I have shown how with different topologies, people may be together 'in place' in ways that include, but are not limited to, the geographic, metric proximity implied in the world 'local'. Being everywhere in the same place, regardless of geometric distance, is part of the deep green environmentalist ideology held by many FoE activists (also see Argyrou 2005). In Chapter 7 I showed how the network topology (Mol and Law 1994) made possible by ICTs generates actual, real presence in the daily lives of FoE activists, with the effect that they are able to generate policies, plans and all sorts of other things of equal value to those produced in situations where presence is in metric proximity. The different topologies depend on what vectors are in play: are the various material actors in place and do they cooperate (is there an infrastructure that can support Internet or international telephone calls); and do the people participating in the practice value or perceive the affordances of presence?

The implication of these different topologies, in regard to questions of globalisation, is that we no longer have to limit ourselves to thinking about the global in opposition to the local. Many other possibilities exist. I am not arguing that these possibilities are the result of so-called forces of globalisation assigned to happenings in the late twentieth century. If, along the lines of ecological phenomenology, the mutual engagements of things in a milieu are understood to constitute place, then a topology will depend on the types of engagement. Different tools offer different effectivities (Ingold 1992), thus the topology generated with email is not the same as that generated by the telephone or in metrically proximate presence, and will be different, too, depending on the ideologies in play. All the more so, we do not need to consider globalisation in terms of the 'evacuation of place' (Giddens 2008) or of a de-localised world (Eriksen 2007), since both such notions distance globalisation and other so-called 'large scale' phenomena from lived experience, when with the ethnography of FoEI, I have shown how both 'the global' and communication across the globe are (inter-)personally and intimately experienced. In other words, globalisation is amenable to phenomenological investigation as much as any other phenomena involving humans. The addition of the ecological ontology makes it possible to understand the nature of humans, especially cognitive involvement and participation in our shared and progressively constituted world. The observations about topologies and presence are only possible thanks to the ethnographic work done with a transnational organisation. And it is precisely because the organisation I worked with is an environmentalist one that the globe is experienced as a single, unfractured and present place.

Taking a step back and looking at the chapters presented in this book at a slight distance, what emerges is a swirling ocean, whose eddies, flows and denser currents—which, viewed from afar, appear stabilised into things—are comprised of different vectors. Some move alongside each other and solidify into human persons: as FoE activists, these persons are made by and make the landscapes, people and activities they go along with. Other vectors, or agglomerates of vectors, come up against one another. At times, these collisions are between forces of equal effect. At other times, humans come up against institutions (composed themselves of transiently stable, eddying swirls of offices, documents, meetings, symbols, memories and conversations) that appear solid and have the effect of solid, supra-personal entities. The solidity derives from distance, which can be due to a lack of familiarity or to a metric distance, in the same way that ocean waves can appear transfixed when seen from an airplane, or the way that water is actually hard if you fall from high enough or at certain angles. With proximity or familiarity the vectors that create solidity can be either reduced or avoided (the EU bureaucracy can be navigated) or they can be deployed deliberately, such as when FoEI is presented to potential funding bodies as a bounded coherent entity.

FoEI activists want to improve the world. They want to see humans living in harmony with each other and with their environments. I have emphasised throughout the book that I share this ideal more than a little with the other activists I worked with, having been a FoE activist myself since 2003. In fact, I joined FoE Malta because I felt more in tune with their principles than with other ENGOs that I had worked with in my previous research (Gatt 2001). However, anthropological work can offer a different take on the questions and challenges that FoE activists tackle. Anthropological approaches can contribute to the activists' hopeful vision without following exactly the same path. This book, therefore, *is* activism. To recognise that the different claims of the world all need to be taken seriously—and in so doing, to point out the limits of FoEI's experiment in inclusive democracy (Chapter 8)—is also to enact an ideology of inclusiveness without having to reduce diversity. In order to find resolutions to the disputes that FoEI activists initiate or participate in, and that will be satisfactory to all the parties concerned, their understanding (and, therefore, their choices and actions) in the world must be allowed equal reality and consequence. Were only one view (whether the environmentalists', the sceptic's or the industrialists') granted purchase on the real world, then this would amount to imperialism of one sort or another. This is the way of 'Science' (Latour 2003), which, by treating other ontologies as mere beliefs, neutralises any challenge they might pose to the ontology of 'the West' (Willerslev 2004). The relativist or social constructivist alternatives, no matter how mainstream, grant 'Science' exclusive access to the real, material world, while relegating everything else to a homogenous category of belief in a 'social' domain detached from the 'material' (Latour 2003; Ingold 1993; 1992, 2000b). Moreover, they offer no possibility for arguing why, what or how some positions may be preferable to others in specific situations. In other words, there is no real basis for negotiation. It is not all a matter of friction, as Tsing (2005) describes collaboration, where alliances are formed with no agreement between the parties. FoEI activists, like anthropologists, endeavour to develop mutual understanding by trying different forms of interaction and communication. They talk using different shapes and sizes and types of discussions—they have staff exchanges, ICTs, marketplaces, mysticas and so on.

I offer this book as a stimulus for further exploration and experimentation in the service of mutual understanding and respect. This is an experiment, on my part, in doing an anthropology *with* Friends of the Earth International.

References

Abélès, Marc. 2004. 'Identity and Borders: An Anthropological Approach to EU Institutions', Twenty-First Century Papers: On-Line Working Papers, University of Wisconsin. www4.uwm.edu/c21/pdfs/workingpapers/abeles Accessed 20 March 2009.

Abu Lughod, Lila. 2000. *Veiled Sentiments: Honor and Poetry in a Bedouin Society. Berkely*. Los Angeles; London: University of California Press.

Adams, Cristina. 2003. 'Pitfalls of Synchronicity: A Case Study of the Caiçaras in the Atlantic Rainforest of South-Eastern Brazil', in *Ethnographies of Conservation: Environmentalism and the Distribution of Privilege*. David Anderson and Eeva Berglund, eds. New York; Oxford: Berghahn Books.

Agre, Phil. 1999. 'Life After Cyberspace'. *Trends in Law Library Management and Technology*. Vol. 10 (1): 1–3.

Allende, Isabel. 2004. *Il Mio Paese Inventato*. Milano: Feltrinelli.

Amit, Vered, ed. 2000. *Constructing the Field*. Canada: Routledge.

Anderson, Benedict. 2006 [1983]. *Imagined Communities: Reflections on the Origin and Spread of Nationalism*. London; New York: Verso.

Anderson, David and Berglund, Eeva, eds. 2003. *Ethnographies of Conservation: Environmentalism and the Distribution of Privilege*. New York; Oxford: Berghahn Books.

Appadurai, Arjun. 2005 [1996]. *Modernity at Large: Cultural Dimensions of Globalization*: Minneapolis; London: University of Minnesota Press.

Argyrou, Vassos. 1997. ' "Keep Cyprus Clean": Littering, Pollution, and Otherness'. *Cultural Anthropology*. Vol. 12 (2): 159–178.

———. 2005. *The Logic of Environmentalism: Anthropology, Ecology and Postcoloniality*. New York; Oxford: Berghahn Books.

Asad, Talal. 1979. 'Anthropology and the Analysis of Ideology'. *Man*, New Series. Vol. 14 (4): 607–627.

———. 1983. 'Anthropological Conceptions of Religion: Reflections on Geertz'. *Man*, New Series. Vol. 18 (2): 237–259.

———. 1987. 'On Ritual and Discipline in Medieval Christian Monasticism'. *Economy and Society*. Vol. 16 (2): 156–203.

Bailey, Frederick George. 1969. *Stratagems and Spoils: A Social Anthropology of Politics*. Oxford: Basil Blackwell.

Bamford, Sandra and Leach, James. 2009. 'Introduction: Pedigrees of Knowledge: Anthropology and the Genealogical Model', in *Kinship and Beyond: The Genealogical Model Reconsidered*. Sandra Bamford and James Leach, eds. New York; Oxford: Berghahn Books.

Barth, Frederik. 2000. 'Boundaries and Connections', in *Signifying Identities: Anthropological Perspectives on Boundaries and Contested values*. Anthony Cohen, ed. London; New York: Routledge.

Bateson, Gregory. 1973. *Steps to an Ecology of Mind*. London: Fontana.

Batteau, Allen. 2001. 'Negations and Ambiguities in the Cultures of Organization'. *American Anthropologist*. Vol. 102 (4): 726–740.

Baumann, Zygmunt. 1995. *Life in Fragments: Essays in Postmodern Morality*. Oxford: Blackwell.

Beck, Ulrich. 1992. *Risk Society: Towards a New Modernity*. London: Sage.

———. 2007 [1994]. 'The Reinvention of Politics: Towards a Theory of Reflexive Modernization', in *Reflexive Modernization: Politics, Tradition and Aesthetics in the Modern Social Order*. Ulirich Beck, Anthony Giddens and Scot Lash, eds. Stanford: Stanford University Press.

Bell, Sandra and Coleman, Simon. 1999. 'The Anthropology of Friendship: Enduring Themes and Future Possibilities', in *The Anthropology of Friendship*. Sandra Bell and Simon Coleman, eds. Oxford; New York: Berg.

Berger, Thomas and Luckmann, Peter. 1971. *The Social Construction of Reality: A Treatise in the Sociology of Knowledge*. Harmondsworth: Penguin Books.

Berglund, Eeva. 1998. *Knowing Science, Knowing Nature*. Cambridge: The White Horse Press.

———. 2001. 'Self-Defeating Environmentalism?' *Critique of Anthropology*. Vol. 21: 317–336.

Bernal, Victoria and Grewal, Inderpal, eds. 2014. *Theorizing NGOS: States, Feminisms and Neoliberalism*. Durham; London: Duke University Press

Biernacki, Richard and Jordan, Jennifer. 2002. 'The Place of Space in the Study of the Social' in *The Social in Question: New Bearings in History and the Social Sciences*. P. Joyce, ed. Canada: Routledge.

Bird-David, Nurit. 1990. 'The Giving Environment: Another Perspective on the Economic System of Gatherer—Hunters'. *Current Anthropology*. Vol. 31 (2): 189–196.

———. 1994. 'Sociality and Immediacy: Or, Past and Present Conversations on Bands'. *Man*, New Series. Vol. 29 (3): 583–603.

———. 1999. 'Animism revisited: Personhood, Environment and Relational Epistemology'. *Current Anthropology*. Vol. 40 (S1): 67–91.

Blaser, Mario. 2010. *Storytelling Globalization From the Chaco and Beyond*. Durham; London: Duke University Press.

———. nd 'On the Properly Political (Disposition for the) Anthropocene'. *Anthropological Theory*.

Boellstorf, Tom. 2008. *Coming of Age in Second life: An Anthropologist Explores the Virtually Human*. Princeton, NJ: Princeton University Press.

Boissevain, Jeremy. 1974. *Friends of Friends: Networks, Manipulators and Coalitions*. Oxford: Basil Blackwell.

———. 1977. 'Of men and Marbles', in *A House Divided: Anthropological Studies of Factionalism*. Richard Salisbury and Marilyn Silverman, eds. St. John's (Nfld): Memorial University of Newfoundland.

———. 1979. 'Network Analysis: A Reappraisal'. *Current Anthropology*. Vol. 20 (2): 392–394.

———. 2004. 'Hotels, Tuna Pens, and Civil Society: Contesting the Foreshore in Malta', in *Contesting the Foreshore: Tourism, Society and Politics on the Coast*. Jeremy Boissevain and Tom Selwyn, eds. Amsterdam: Amsterdam University Press.

Boissevain, Jeremy and Gatt, Caroline. 2011 'Environmentalists in Malta: The Growing Voice of Civil Society', in *Contested Mediterranean Spaces: Ethnographic Essays in Honour of Charles Tilly*. Mario Kousis, Tom Selwyn and David Clark, eds. New York; Oxford: Berghahn Books.

Boissevain, Jeremy and Theuma, Nadia. 1998. 'Contested Space: Planners, Tourists, Developers and Environmentalists in Malta', in *Anthropological Perspectives on Local Development*. Simone Abram and Jacqueline Waldren, eds. London: Routledge.

Bond, Patrick. 2015. 'Can Climate Activists "Movement Below" Transcend Negotiators "Paralysis Above"'. *Journal of World-Systems Research*. Vol. 21 (2): 250–270.

Borg Barthet, Justin. 2010. 'A New Approach to the Governing of Companies in the EU: A Legislative Proposal'. *Journal of Private International Law*. Vol. 6 (3): 589–621.

Bourdieu, Pierre. 1984. *Homo Academicus*. Stanford: Stanford University Press.

———. 1985. 'The Social Space and the Genesis of Groups'. *Theory and Society*. Vol. 14 (6): 723–744.

———. 1990. *The Logic of Practice*. Stanford, CA: Stanford University Press.

Briggs, Charles. 1996. 'The Politics of Discursive Authority in Research on the "Invention of Tradition"'. *Cultural Anthropology*. Vol. 11 (4): 435–469.

Brosius, Peter. 1999. 'Analyses and Interventions: Anthropological Engagements With Environmentalism'. *Current Anthropology*. Vol. 40 (3): 227–309.

Brubaker, Roger and Cooper, Frederick. 2000. 'Beyond "Identity"'. *Theory and Society*. Vol. 29 (1): 1–47.

Bullard, Robert. 1994. *Dumping in Dixie*. Boulder: Westview Press.

Calhoun, Craig. 1991. 'Indirect Relationships and Imagined Communities: Large-Scale Social Integration and the Transformation of Everyday Life', in *Social Theory for a Changing Society*. Pierre Bourdieu and James Coleman, eds. New York: Russell Sage Foundation.

———. 1995. 'Community Without Propinquity Revisited: Communications Technology and the Transformation of the Urban Public Sphere'. *Sociological Inquiry*. Vol. 68 (3): 373–397.

Campbell, Howard. 1993. 'Tradition and the New Social Movements: The Politics of Isthmus Zapotec Culture'. *Latin American Perspectives*. Vol. 20 (3): 83–97.

Candea, Matei. 2007. 'Arbitrary Locations: In Defence of the Bounded Field-site'. *Journal of the Royal Anthropological Institute*. Vol. 13: 167–184.

———. 2010. 'Endo/Exo'. *Common Knowledge*. Vol. 17 (1): 146–150.

Capra, Fritjof. 1975. *The Tao of Physics*. Boston: Flamingo Press.

Carruthers, Mary. 1998. *The Craft of Thought: Meditation, Rhetoric, and the Making of Images*. Australia: Cambridge University Press.

Castells, Manuel. 1996. *The Rise of the Network Society*. Oxford: Blackwell.

———. 1997. *The Power of Identity*. Oxford: Blackwell.

Chalmers, David. 1996. *The Conscious Mind: In Search of a Fundamental Theory*. New York: Oxford University Press.

Choy, Timothy. 2005. 'Articulated Knowledges: Environmental Forms After Universality's Demise'. *American Anthropologist*. Vol. 107: 5–18.

Chua, Liana and Mathur, Nayanika. n.d. 'Introducing the 'We' in and of Anthropology', in *Who Are "We"? Reimagining Alterity and Affinity in Anthropology*. Liana Chua and Nayanika Mathur, eds. New York; London: Berghahn Books.

Clark, Andy. 2001. *Mindware: An Introduction to the Philosophy of Cognitive Science*. Oxford; New York: Oxford University Press.

———. 2008. *Supersizing the Mind: Embodiment, Action and Cognitive Extension*. Oxford; New York: Oxford University Press.

Clough, Paul. 2001. 'Conclusion: The Political Economy Behind the Powers of Good and Evil', in *Powers of Good and Evil: Moralities, Commodities and Popular Belief*. Paul Clough and Jon Mitchell, eds. New York; Oxford: Berghahn Books.

Cohen, Anthony. 2000. 'Introduction Discriminating Relations: Identity, Boundary and Authenticity', in *Signifying Identities: Anthropological Perspectives on Boundaries and Contested Values*. Anthony Cohen, ed. London; New York: Routledge.

Coleman, Simon and Collins, Peter. 2006. 'Introduction: 'Being . . . Where?' Performing Fields on Shifting Grounds', in *Locating the Field: Space, Place and Context in Anthropology*. S. Coleman and P. Collins, eds. Oxford; New York: Berg.

Collins, Peter. 2008. 'The Practice of Discipline and the Discipline of Practice', in *Exploring Regimes of Discipline: The Dynamics of Restraint*. Noel Dyck, ed. Oxford; New York: Berghahn Books.

Connolly, John and Prothero, Andrea. 2008. 'Green Consumption: Life-politics, Risk and Contradictions'. *Journal of Consumer Culture*. Vol. 8 (1): 117–145.

Correll, Shelley. 1995. 'The Ethnography of an Electronic Bar: The Lesbian Cafe'. *Journal of Contemporary Ethnography*. Vol. 24 (3): 270–298.

Croll, Elizabeth and Parkin, David, eds. 1992. *Bush Base: Forest Farm Culture, Environment and Development*. London; New York: Routledge.

Dandelion, Pink and Collins, Peter. 2008. 'Introduction', in *The Quaker Condition: The Sociology of a Liberal Religion*. Pink Dandelion and Peter Collins, eds. Cambridge: Cambridge Scholars Publishing.

Deeb, Hadi Nicholas and Marcus, George E. 2011. 'In the Green Room: An Experiment in Ethnographic Method at the WTO'. *Political and Legal Review*. Vol. 34 (1): 51–76.

Di Chiro, Giovanna. 1996. 'Nature as Community: The Convergence of Environment and Social Justice', in *Uncommon Ground: Toward Reinventing Nature*. William Cronon, ed. New York: W.W. Norton & Co.

Dirlik, Arif. 1997. *The Post-Colonial Aura: Third World Criticism in the Age of Global Capitalism*. Coulder: Westview Press.

Doherty, Brian. 2006. 'Friends of the Earth International: Negotiating a North-South Identity'. *Environmental Politics*. Vol. 15 (5): 860–880.

Doherty, Brian and Doyle, Timothy. 2009. ' "We Are Heavily in Solidarity in this Room": Accountability, Representation and Democracy in Friends of the Earth International'. Paper for ECPR Joint Sessions, Lisbon, 14–19 April.

Douglas, Mary and Wildavsky, Aaron. 1983. *Risk and Culture: An essay on the Selection of Technological and Environmental Dangers*. Berkeley; Los Angeles; London: University of California Press.

Drucker, Peter. 1997. 'The Global Economy and the Nation State'. *Foreign Affairs*. Vol. 76 (5): 159–170.

Durkheim, Emile. 1973. 'The Dualism of Human Nature and Its Social Conditions', in *On Morality and Society*. Robert Bellah, ed. London; Chicago: University of Chicago Press.

Durkheim, Émile. 1982. [1st pub. 1895]. Lukes, Steven, ed. *The Rules of Sociological Method and Selected Texts on Sociology and its Method*. W. D. Halls (translator). New York: Free Press.

Dyck, Noel, ed. 2008. *Exploring Regimes of Discipline: The Dynamics of Restraint*: Oxford; New York: Berghahn Books.

Eckersley, Robin. 1989. 'Green Politics and the New Class: Selfishness or Virtue?' *Political Studies*. Vol. 37 (2): 205–223.

Edelman, Marc. 2001. 'Social Movements: Changing Paradigms and Forms of Politics'. *Annual Review of Anthropology*. Vol. 30: 285–317.

———. 2002. 'Toward an Anthropology of Some New Internationalisms: Small Farmers in Global Resistance Movements'. *Focaal*. Vol. 40: 103–122.

Eisenlohr, Patrick. 2011. 'Introduction: What Is a Medium? Theologies, Technologies and Aspirations'. *Social Anthropology*. 19: 90–102.

Engelke, Matthew. 2008. 'Objects of Evidence'. *Journal of the Royal Anthropological Institute* (N.S.). Vol. 14: S1–21.

———. 2011. 'Media, Mediation, Religion. (Debate with Charles Hirschkind)'. *Social Anthropology*. Vol. 19: 90–102.

Eriksen, Thomas Hylland. 1998. *Common Denominators: Ethnicity, Nation-Building and Compromise in Mauritius*. Oxford; New York: Berg.

———. ed. 2003. *Globalisation*. London: Pluto Press.

———. 2006. *Engaging Anthropology: The Case for a Public Presence*. Oxford: Berg.

———. 2007. *Globalization: The Key Concepts*. Oxford; New York: Berg.

Escobar, Arturo. 1994. 'Welcome to Cyberia: Notes on the Anthropology of Cyberculture'. *Current Anthropology*. Vol. 35 (3): 211–231.

———. 1999. 'After Nature: Steps to an Antiessentialist Political Ecology'. *Current Anthropology*. Vol. 40 (1): 1–30.

———. 2008. *Territories of Difference: Place, Movements, Life*, Redes. Durnham: Duke University Press.

Falzon, Mark Anthony. 2004. *Cosmopolitan Connections: The Sindhi Diaspora 1860–2000*. The Netherlands: Brill.

———. 2009. 'Introduction: Multi-Sited Ethnography: Theory, Praxis and Locality in Contemporary Research', in *Multi-Sited Ethnography: Theory, Praxis and Locality in Contemporary Research*. Mark Anthony Falzon, ed. Burlington: Ashgate.

Farnell, Brenda. 1999. 'Moving Bodies: Acting Selves'. *Annual Review of Anthropology*. Vol. 28: 341–373.

———. 2000. 'Getting Out of the Habitus'. *The Journal of the Royal Anthropological Institute*. Vol. 6 (3): 397–418.

Ferguson, James. 1990. *The Anti-Politics Machine: 'Development,' Depoliticization, and Bureaucratic Power in Lesotho*. Cambridge: Cambridge University Press.

Fischer, William. 1997. 'Doing Good? The Politics and Antipolitics of NGO Practices'. *Annual Review of Anthropology*. Vol. 26: 439–469.

Fortun, Kim. 2001. *Advocacy After Bhopal: Environmentalism, Disaster, New Global Orders*. London: University of Chicago Press.

Foucault, Michel. 1988. 'Technologies of the Self', in *Technologies of the Self: A Seminar With Michel Foucault*. Martin, Luther ed. Amherst: The University of Massachusetts Press. *FoEI Link*, 1981–1985, FoEI Archives, Amsterdam.

Foucault, Michel. 1995 [1977]. *Discipline and Punish: The Birth of the Prison*. New York: Vintage Books.

Fraser, Nancy. 2003. 'From Discipline to Flexibilization? Rereading Foucault in the Shadow of Globalization'. *Constellations*. Vol. 10 (2): 160–171.

Garsten, Christina. 1994. *Apple World: Core and Periphery in a Transnational Organizational Culture: A Study of Apple Computer Inc*. Stockholm: Almqvist & Wiksell International

Garsten, Christina and Wulff, Helena. 2003. 'Introduction: From People of the Book to People of the Screen', in *New Technologies at Work: People, Screens and Social Virtuality*. Christina Garsten and Helena Wulff, eds. Oxford; New York: Berg.

Gatt, Caroline. 2001. 'Environmentalism in the Maltese Context: The Case of Nature Trust'. Unpublished BA (Hons) Dissertation, Mediterranean Institute, University of Malta.

———. 2005. 'Eliciation and Consensus-Building as Management Tools', in *Community Centres: Promoting Sustainable Living*. Vincent Caruana, ed. Malta: Outlook Coop.

———. 2009. 'Emplacement in Multi-Sited Practice/Theory', in *Multi-Sited Ethnography: Theory, Praxis and Locality in Contemporary Research*. Mark Anthony Falzon, ed. Burlington: Ashgate.

———. 2010. 'Serial Closure: Generative Reflexivity and Restoring Confidence in/ of Anthropologists', in *Unquiet Pasts: Risk Society, Lived Cultural Heritage and Re-Designing Reflexivity*. Stephanie Keorner and Ian Russel, eds. Burlington: Ashgate.

———. 2011. 'By Way of Theatre: Design Anthropology and the Exploration of Human Possibilities'. Conference Proceedings 'Participatory Innovation Conference', University of Southern Denmark. http://spirewire.sdu.dk/PINC-proceedings-web.pdf

———. 2017 'The Liveliness of Books', in *The Voices of the Pages*. Caroline Gatt, ed. Aberdeen: KFI books.

Geertz, Armin. 1997. 'From Stone Tablets to Flying Saucers: Tradition and Invention in Hopi Prophecy', in *Present is Past: Some Uses of Tradition in Native Societies*. Marie Mauzé, ed. Lanham, MD: University Press of America.

Geertz, Clifford. 1973. *The Interpretation of Cultures*. London: Fontana Press.

———. 1983. *Local Knowledge: Further Essays in Interpretive Anthropology*. London: Fontana Press.

Gibson, James. 1979. *The Ecological Approach to Visual Perception*. Boston: Houghton Mifflin.

Giddens, Anthony. 2007 [1994]. 'Living in a Post-Traditional Society', in *Reflexive Modernization: Politics, Tradition and Aesthetics in the Modern Social Order*. Ulrich Beck, Anthony Giddens and Scott Lash, eds. Cambridge: Polity Press.

———. 2008 [1991]. *Modernity and Self-Identity: Self and Society in the Late Modern Age*. Cambridge: Polity Press.

———. 2009 [1984]. *The Constitution of Society: Outline of the Theory of Structuration*: Cambridge: Polity Press.

Graeber, David. 2007. 'There Never Was a West'. The Anarchist library.org

———. 2013. *The Democracy Project: A History, a Crisis, a Movement*. New York: Spiegel and Grau.

———. 2015. 'Radical Alterity Is Just Another Way of Saying "Reality: A Reply to Edoardo Viveiros de Castro"'. *Hau*. Vol. 5 (2): 1–41.

Gray, Chris and Discroll, Mark. 1992. 'What's Real About Virtual Reality? Anthropology of, and in, Cyberspace'. *Visual Anthropology Review*. Vol. 8 (2): 39–49.

Green, Sarah. 2003. 'Digital Ditches: Working in the Virtual Grass Roots', in *New Technologies at Work: People, Screens and Social Virtuality*. Christina Garsten and Helena Wulff, eds. Oxford; New York: Berg.

Green, Sarah, Harvey, Penny and Knox, Hannah. 2005. 'Scales of Place and Networks: An Ethnography of the Imperative to Connect through Information and Communications Technology'. *Current Anthropology*. Vol. 46 (5): 805–826.

Grove-White, Robin. 1993. 'Environmentalism: A New Moral Discourse for Technological Society?' in *Environmentalism: The View From Anthropology*. Kay Milton, ed. London: Routledge.

Gupta, Akhil. 1995. 'Blurred Boundaries: The Discourse of Corruption, The Culture of Politics, and the Imagined State'. *American Ethnologist*. Vol. 22 (2): 375–402.

Gupta, Akhil and Ferguson, James. 1997. *Anthropological Locations: Boundaries and Grounds of a Field Science*. Berkeley: University of California Press.

Hann, Chris. 1996. 'Introduction: Challenging Western Models', in *Civil Society: Challenging Western Models*. Chris Hann and Elizabeth Dunn, eds. Canada: Routledge.

Hannerz, Ulf. 1996. *Transnational Connections: Culture, People and Places*. Canada: Routledge.

Hanson, Allan. 1997. 'Empirical Anthropology, Postmodernism, and the Invention of Tradition', in *Present is Past: Some Uses of Tradition in Native societies*. Marie Mauzé, ed. Lanham, MD: University Press of America.

Haraway, Donna. 1991. *Simians, Cyborgs and Women: The Reinvention of Nature*. New York: Routledge.

———. 2003. *The Companion Species Manifesto: Dogs, People, and Significant Otherness*. Chicago: Prickly Paradigm Press.

Harkness, Rachel. 2009. 'Thinking Building Dwelling: Examining Earthships in Taos and Fife'. PhD thesis, University of Aberdeen.

Harvey, David. 2000. *Spaces of Hope*. Edinburgh: Edinburgh University Press.

Hasselstrom, Anna. 2003. 'Real-Time, Real-Place Market: Transnational Connections and Disconnections in Financial Markets', in *New Technologies at Work: People, Screens and Social Virtuality*. Christina Garsten and Helena Wulff, eds. Oxford; New York: Berg.

Hastrup, Kirsten and Hervik, Peter. 1994. 'Introduction', in *Social Experience and Anthropological Knowledge*. Kirsten Hastrup and Peter Hervik, eds. London; New York: Routledge.

Hatch, Elvin. 1973. *Theories of Man and Culture*. New York: Columbia University Press.

Hendry, Joy and Watson, C.W. 2001. 'Introduction', in *An Anthropology of Indirect Communication*. Joy Hendry and C.W. Watson, eds. London; New York: Routledge.

Herzfeld, Michael. 1985. *The Poetics of Manhood: Contest and Identity in a Cretan Mountain Village*. Princeton, NJ: Princeton University Press.

———. 2001. *Anthropology: Theoretical Practice in Culture and Society*. London: Blackwell Publishers Limited.

Hilhorst, Dorothea. 2003. *The Real World of NGOs*. London; New York: Zed Books Ltd.

Hirsch, Eric and Stewart, Charles. 2005. 'Introduction: Ethnographies of Historicity'. *History and Anthropology*. Vol. 16 (3): 261–274.

Hobsbawm, Eric. 1983. 'Introduction: Inventing Traditions', in *The Invention of Tradition*. Eric Hobsbawm and Thomas Ranger, eds. Cambridge; New York: Cambridge University Press.

Hodges, Matthew. 2008. 'Rethinking Time's Arrow: Bergson, Deleuze and the Anthropology of Time'. *Anthropological Theory*. Vol. 8 (4): 399–429.

Holmes, Douglas and Marcus, George. 2008. 'Collaboration Today and the Reimagination of the Classic Scene of Fieldwork Encounter'. *Collaborative Anthropologies*. Vol. 1: 81–101.

Holmes, George. 2011. 'Conservation's Friends in High Places: Neoliberalism, Networks, and the Transnational Conservation Elite'. *Global Environmental Politics*. Vol. 11 (4): 1–21.

Hopgood, Stephen. 2006. *Keepers of the Flame: Understanding Amnesty International*. Ithaca, NY; London: Cornell University Press.

Horst, Heather and Miller, Daniel. 2006. *The Cell Phone: An Anthropology of Communication*. Berg Publications, New York, NY, USA.

Horton, Dave. 2003. 'Green Distinctions: The Performance of Identity Among Environmental Activists'. *The Sociological Review*. Vol. 51 (Supplement 2): 63–77.

Huxley, Aldous. 1994 [1954]. *The Doors of Perception*. London: Flamingo.

Ietto-Gillies, Grazia. 2005. *Transnational Corporations and International Production: Concepts, Theories and Effects*. Cheltenham, UK: Edward Elgar Publishing Limited.

Inda, Jonathan and Rosaldo, Renato. 2002. 'Tracking Global Flows', in *The Anthropology of Globalization: A Reader*. Javier Inda and Renato Rosaldo, eds. Oxford: Blackwell.

Ingold, Tim. 1987. *The Appropriation of Nature: Essays on Human Ecology and Social Relations*. Iowa City: University of Iowa Press.

———. 1992. 'Culture and the Perception of the Environment', in *Bush Base: Forest Farm Culture, Environment and Development*. Elizabeth Croll and David Parkin, eds. London; New York: Routledge.

———. 1993. 'The Art of Translation in a Continuous World', in *Beyond Boundaries: Understanding, Translation and Anthropological Discourse*. Gisli Pálsson, ed. Oxford: Berg.

———. 1996. 'The Concept of Society Is Theoretically Obsolete', in *Key Debates in Anthropology*. Tim Ingold, ed. London: Routledge.

———. 1999. 'Human Nature and Science'. *Interdisciplinary Science Review*. Vol. 24 (4): 250–254.

———. 2000a. *The Perception of the Environment: Essays in Livelihood, Dwelling and Skill*. Canada: Routledge.

———. 2000b. 'Concluding Comment', in *Negotiating Nature: Culture, Power and Environmental Argument*. Alf Hornborg and Gisli Palsson, eds. Lund: Lund University Press.

———. 2001. 'From the Transmission of Representations to the Education of Attention', in *The Debated Mind: Evolutionary Psychology Versus Ethnography*. Harvey Whitehouse, ed. Oxford: Berg.

———. 2002. 'Communication and Communion'. *Behavioral and Brain Sciences*. Vol. 25 (5): 627–628.

———. 2005a. 'Epilogue: Towards a Politics of Dwelling'. *Conservation and Society*. Vol. 3 (2): 501–508.

———. 2005b. 'Movement, Knowledge and Description', in *Holistic Anthropology: Emergence and Convergence*. D. Parkin and S. Ulijaszek, eds. New York; Oxford: Berghahn Books.

———. 2006. 'Rethinking the Animate, Re-Animating Thought'. *Ethnos*. Vol. 71 (1): 9–20.

———. 2007a. *Lines: A Brief History*. London; New York: Routledge.

———. 2007b. 'Material Against Materiality'. *Archaeological Dialogues*. Vol. 14: 1–16.

———. 2007c. 'Anthropology is *not* Ethnography'. Radcliffe-Brown Lecture at the British Academy.

———. 2008a. 'Bindings Against Boundaries: Entanglements of Life in an Open World'. *Environment and Planning A*. Vol. 40 (8): 1796–1810.

———. 2008b. 'When ANT Meets SPIDER: Social Theory for Arthropods', in *Material Agency: Towards a Non-Anthropocentric Approach*. Carl Knappett and Lambros Malafouris, eds. New York: Springer.

———. 2009. 'No More Ancient; No More Human: The Future Past of Archaeology and Anthropology'. Keynote speech presented to the Conference of the Association of Social Anthropologists of the UK and Commonwealth. University of Bristol, 6–9 April.

———. 2010. 'Ways of Mind-Walking: Reading, Writing, Painting'. *Visual Studies*. Vol. 25 (1): 15–23.

———. 2011a. *Being Alive: Essays on Movement, Knowledge and Description*. London; New York: Routledge.

———. 2011b. 'Becoming and Its Limits'. Paper presented at the Durham Senior Seminars, 9 March.

———. 2013. *Making: Anthropology, Archaeology, Art and Architecture*. London; New York: Routledge.

———. 2015. *The Life of Lines*. London; New York: Routledge.

Ingold, Tim and Hallam, Liz. 2007. 'Creativity and Cultural Improvisation: An Introduction', in *Creativity and Cultural Improvisation*. Tim Ingold and Liz Hallam, eds. Oxford; New York: Berg.

Ingold, Tim, Riches, David and Woodburn, James. 1988. *Hunters and Gatherers Vol 1: Hisotry, Evolution and Social Change*. Oxford, New York: Berg/Bloomsbury.

Jackson, Michael. 1983. 'Knowledge of the Body'. *Man*, New Series. Vol. 18 (2): 327–345.

Jackson, Michael. 2002. 'Familiar and foreign bodies: a phenomenological exploration of the human-technology interface'. *Journal of the Royal Anthropological Institute*. Vol. 8 (2), 333–346.

James, Wendy and Aston, Judith. 2006. 'Social Sounds: Collaborative Rhythms in Work, Music and Language among Uduk Speaking People (Sudan/Ethiopia)'. Paper presented at the Conference: Sound and Anthropology: Body, Environment and Human Sound Making, St. Andrews University, 19–21 June.

Järpe, Anna. 2007. 'Ever Against the Wind . . . ' Lifescapes and Environmental Perception Among Sámi Reindeer Tenders in Västerbotten, Sweden". PhD thesis, University of Aberdeen.

Jiménez, Alberto. 2007. 'Introduction', in *The Anthropology of Organisations*. Alberto Jiménez, ed. Burlington: Ashgate Publishing Limited.

Kendrick, Justin. 2009. 'The Paradox of Indigenous People's Rights'. *World Anthropologies Journal*. Vol. 4: 11–45.

Knappett, Carl and Malafouris, Lambros. 2008. 'Material and Non-Human Agency: An Introduction', in *Material Agency: Towards a Non-Anthropocentric Approach*. Carl Knappett and Lambros Malafouris, eds. New York: Springer.

Knorr Cetina, Karin D. and Bruegger, Urs. 2000. 'The Market as an Object of Attachment: Exploring Postsocial Relations in Financial Markets'. *Canadian Journal of Sociology*. Vol. 25 (2): 141–168.

Kohn, Tamara. 2008. 'Creatively Sculpting the Self Through the Discipline of Martial Arts Training', in *Exploring Regimes of Discipline: The Dynamics of Restraint*. Noel Dyck, ed. Oxford; New York: Berghahn Books.

Kusch, Martin. 1991. *Foucault's Strata and Fields*: Dordrecht: Kluwer Academic Publishers.

Larsen, Peter Bille. 2013. 'The Politics of Technicality: Guidance Culture in Environmental Governance and the International Sphere', in *The gloss of Harmony: The politics of policy-making in Multilateral organisations*. Müller Birgit, ed. London; New York: Pluto Press.

———. 2016. 'The Good, the Ugly and the Dirty Harry's of Conservation: Rethinking the Anthropology of Conservations NGOs'. *Conservation and Society*. Vol. 14 (1): 21–33.

Latour, Bruno. 1987. *Science in Action: How to Follow Scientists and Engineers Through Society*. Cambridge, MA: Harvard University Press.

———. 1993. *We Have Never Been Modern*, trans Catherine Porter. New York: Harvester Wheatsheaf.

———. 2003. *Politics of Nature: How to Bring the Sciences into Politics*. Cambridge, MA: Harvard University Press.

———. 2005a. 'From Realpolitik to Dingpolitk or How to Make Things Public', in *Making Things Public: Atmopheres of Democracy*. Bruno Latour and Peter Weibel, eds. Cambridge, MA: MIT Press.

———. 2005b. *Reassembling the Social: An Introduction to Actor-Network Theory*. Oxford: Oxford University Press.

———. 1987. *Science in Action: How to Follow Scientists and Engineers through Society*. Harvard: Harvard University Press.

Lave, Jean. 1988. *Cognition in Practice: Mind, Mathematics and Culture in Everyday Life*. Cambridge: Cambridge University Press.

Law, John. 1992. 'Notes on the Theory of the Actor Network: Ordering, Strategy and Heterogeneity'. published by the Centre for Science Studies, Lancaster University. www.comp.lancs.ac.uk/sociology/papers/Law-Notes-on-ANT.pdf

Leach, James. 2006. 'Out of Proportion? Anthropological Description of Power, Regeneration and Scale on the Rai Coast of Papua New Guinea', in *Locating the Field: Space, Place and Context in Anthropology*. Simon Coleman and Peter Collins, eds. Oxford; New York: Berg.

———. 2010. 'Intervening With the Social: Ethnographic Practice and Tarde's Image of Relations Between Subjects', in *The Social After Gabriel Tarde*. Matei Candea, ed. London: Routledge.

Lee, Jo and Ingold, Tim. 2006. 'Fieldwork on Foot: Perceiving, Routing, Socializing', in *Locating the Field: Space, Place and Context in Anthropology*. Simon Coleman and Peter Collins, eds. Oxford; New York: Berg.

Leeder, Kim. 2007. 'Technology and Communication in the Environmental Movement'. *Electronic Green Journal*. 25. http://escholarship.org/uc/item/9gt4h74z Accessed 9 February 2009

Lefebvre, Henri. 1991. *The Production of Space*, tran. Donald Nicholson-Smith. Oxford; Cambridge, MA: Blackwell.

Legat, Allice. 2007. 'Walking the Land, Feeding the Fire: Becoming and Being Knowledgeable Among the Tlicho Dene', PhD thesis, University of Aberdeen.

Lincoln, Amber. 2011. 'Body Techniques of Health: Making Products and Shaping Selves in Northwest Alaska'. *Etudes/Inuit/Studies*. Vol. 34 (2): 39–50.

Linde, Charlotte. 1993. *Life Stories: The Creation of Coherence*. New York; Oxford: Oxford University Press.

Linnekin, Jocelyn. 1992. 'On the Theory and Politics of Cultural Construction in the Pacific'. *Oceania*. Vol. 62 (4): 249–263.

Lister, Sarah. 2003. 'NGO Legitimacy: Technical Issue or Social Construct?' *Critique of Anthropology*. Vol. 23 (2): 175–192.

Little, Paul. 1999. 'Environments and Environmentalisms in Anthropological Research: Facing a New Millennium'. *Annual Review of Anthropology*. Vol. 28: 253–284.

Loizos, Peter and Paptaxiarchis, Evthymios. 1991. 'Introduction: Gender and Kinship in Marriage and Alternative Contexts', in *Contested Identities: Gender and Kinship in Modern Greece*. Peter Loizos and Evthymios Papataxiarchis, eds. Princeton, NJ: Princeton University Press.

Lund, Katrín. 2005. 'Finding Place in Nature: 'Intellectual' and Local Knowledge in a Spanish Natural Park'. *Conservation and Society*. Vol. 3 (2): 371–387.

Macnaghten, Phil. 2003. 'Embodying the Environment in Everyday Life Practices'. *The Sociological Review*. Vol. 51 (1): 63–84.

Malinowski, Bronislaw. 1922. *Argonauts of the Western Pacific*. London; New York: Routledge.

Marcus, George. 1995. 'Ethnography in/of the World System: The Emergence of Multi-Sited Ethnography'. *Annual Review of Anthropology*. Vol. 24: 95–117.

———. 1998. *Ethnography Through Thick and Thin*. Princeton, NJ: Princeton University Press.

———. 2001. 'From Rapport Under Erasure to Theaters of Complicit Reflexivity'. *Qualitative Inquiry*. Vol. 7 (4): 519–528.

———. 2007. 'How Short Can Fieldwork Be?' *Social Anthropology/Anthropologie Sociale*. Vol. 15 (3): 353–357.

———. 2009. 'The Green Room, Off-Stage: In Site-Specific Performance Art and: Ethnographic Encounters'. Paper presented at Performance, Art et Anthropologie, Colloque International, Paris, 11–12 March.

Marcus, George and Fischer, Michael. 1999 [1986]. *Anthropology as Cultural Critique: An Experimental Moment in the Human Sciences*. Chicago; London: University of Chicago Press.

Martin, Jay. 1993. *Force Fields: Between Intellectual History and Cultural Critique*. New York; London: Routledge.

Marvin, Carolyn. 1988. *When Old Technologies Were New*. New York; Oxford: Oxford University Press.

Mauzé, Marie. 1997. 'On Concepts of Tradition: An Introduction', in *Present Is Past: Some Uses of Tradition in Native Societies*. Marie Mauzé, ed. Lanham, MD: University Press of America.

McGill, Kenneth. 2016. *Global Inequality*. Toronto: University of Toronto Press.

Meyer, John and Jeppersen, Ronald. 2000. 'The "Actors" of Modern Society: The Cultural Construction of Social Agency'. *Sociological Theory*. Vol. 18 (1): 100–120.

Miller, Daniel and Slater, Don. 2000. *The Internet: An Ethnographic Approach*. Oxford; New York: Berg.

Milton, Kay. 1993. 'Introduction: Environmentalism and Anthropology', in *Environmentalism: The View From Anthropology*. Kay Milton, ed. London: Routledge.

———. 1996. *Environmentalism and Cultural Theory: Exploring the Role of Anthropology in Environmental Discourses*. London: Routledge.

———. 2002. *Loving Nature: Towards an Ecology of Emotion*. London: Routledge.

Mitchell, Jon. 2002. *Ambivalent Europeans: Ritual, Memory and the Public Sphere in Malta*. London: Routledge.

Miyazaki, Hirokazu. 2004. *The Method of Hope: Anthropology, Philosophy, and Fijian Knowledge*. Stanford, CA: Stanford University Press.

Moisander, Johanna and Pesonen, Sinnika. 2002. 'Narratives of Sustainable Ways of Living: Constructing the Self and the Other as a Green Consumer'. *Management Decision*. Vol. 40 (4): 329–342.

Mol, Annemarie and Law, John. 1994. 'Regions, Networks and Fluids: Anaemia and Social Topology'. *Social Studies of Sciences*. Vol. 24 (4): 641–671.

Morse, Margaret. 1998. *Virtualities: Television, Media Art and Cyberculture*. Indiana: Indiana University Press.

Mosse, David. 2005. *Cultivating Development: An Ethnography of Aid Policy and Practice*. London; Ann Arbor, MI: Pluto Press.

———. 2007. 'Is Good Policy Unimplementable? Reflections on the Ethnography of Aid Policy', in *The Anthropology of Organisations*. A. Jiménez, ed. Burlington: Ashgate Publishing Limited.

Mosse, David and Lewis, David. 2006. 'Encountering Order and Disjuncture: Contemporary Anthropological Perspectives on the Organization of Development'. *Oxford Development Studies*. Vol. 34 (1): 1–13.

Müller, Birgit. 2011. 'The Elephant in the Room: Multistakeholder Dialogue on Agricultural Biotechnology in the Food and Agriculture Organization', in *Policy Worlds: Anthropology and the Analysis of Contemporary Power*. Cris Shore, Susan Wright and Davide Però. eds. New York; Oxford: Berghahn Books.

———. 2013. 'Introduction: Lifting the Veil of Harmony: Anthropologists Approach International Organisations', in *The Gloss of Harmony: The Politics of Policy-Making in Multilateral Organisations*. Birgit Müller, ed. London; New York: Pluto Press.

Murphy, Michelle. 2010. 'Anticipation'. Paper given at the conference in Making and Opening: Entangling Design and Social Science, Goldsmiths, University of London, 24 September.

Neveling, Patrick. 2010. 'Some Remarks on the Scalar Structurations of Capitalism and the Anthropology of the Twentieth Century Global System'. Paper presented at EASA, Maynooth, 24–28 August.

Nustad, Knut. 2003. 'Considering Global/Local Relations: Beyond Dualism', in *Globalisation*. Thomas Hyllans Eriksen, ed. New York: Pluto Press.

Ochs, Elinor and Capps, Lisa 1996. 'Narrating the Self'. *Annual Review of Anthropology*. Vol. 25: 19–43.

Okely, Judith. 1992. 'Anthropology and Autobiography: Participatory Experience and Embodied Knowledge', in *Anthropology & Autobiography*. Judith Okely and Helen Callaway, eds. Canada: Routledge.

———. 1994. 'Thinking Through Fieldwork', in *Analyzing Qualitative Data*. Anthony Bryman and Richard G. Burgess, eds. London: Routledge.

———. 1996. *Own or Other Culture*. London: Routledge.

———. 2001. 'Visualism and Landscape: Looking and Seeing in Normandy'. *Ethnos*. Vol. 66 (1): 99–120.

Okely, Judith and Callaway, Helen. 1992. 'Preface', in *Anthropology & Autobiography*. Judith Okely and Helen Callaway, eds. Canada: Routledge.

Orr, Yancey, Lansing, Stephen and Dove, Michael. 2015. 'Environmental Anthropology: Systemic Perspectives'. *Annual Review of Anthropology*. Vol. 44: 153–168.

Ortner, Sherry. 1984. 'Theory in Anthropology since the Sixties', in *Comparative Studies in Society and History*, Vol 26 (1): 126-166.

Peacock, Vita. 2013 'Agency and the Anstoß: Max Planck Directors as Fichtean Subjects', Special Issue on the Study of Organisations (ed. V. Peacock). *Anthropology in Action*. Vol. 20 (2): 6–16.

Pfaffenberger, Bryan. 1992. 'Social Anthropology of Technology'. *Annual Review of Anthropology*. Vol. 21: 491–516.

Pickerill, Jenny. 2001. ' "Environmentalists" Internet Activism in Britain'. *Peace Review*. Vol. 13 (3): 365–370. www.jennypickerill.info/pubs.html Accessed 10 February 2011.

Plüs, Caroline. 1995. 'A Sociological Analysis of the Modern Quaker Movement'. PhD thesis, University of Oxford.

Porath, Nathan. 2007. 'Being Human in a Dualistically-Conceived Embodied World: Descartes' Dualism and Sakais' Universalist Concepts of (Altered) Consciousness, Inner-Knowledge and Self', in *Anthropology and Science: Epistemologies in Practice*. Jeanette Edwards, Penny Harvey and Peter Wade, eds. Oxford; New York: Berg.

Porrit, Jonathan. 1991. *Save the Earth*. London: Dorling Kindersley Publishers Ltd.

Rabinow, Paul. 1984. *The Foucault Reader*. London: Penguin.

Ram, Kalpana. 2006. 'A New Consciousness Must Come': Affectivity and Movement in Tamil Dalit Women's Activist Engagement With Cosmopolitan Modernity', in *Anthropology and the New Cosmopolitanism: Rooted, Feminist and Vernacular Perspectives*. Pnina Werbner, ed. Oxford; New York: Berg.

Ranger, Terence. 1993, 'The Invention of Tradition Revisited', in *Legitimacy and the State in Twentieth Century Africa: Essays in Honour of A. H. M. Kirk-Greene*. Terence Ranger and Olufemi Vaughan, eds. Houndmills, Basingstoke, Hampshire: Palgrave Macmillan, in association with St. Antony's College, Oxford.

Rapport, Nigel. 2002. Part II the Truth of Movement, the Truth as Movement: Post-Cultural Anthropology and Narrational Identity", in *The Trouble With Community: Anthropological Reflections on Movement, Identity and Collectivity*. Vered Amit and Nigel Rapport. London: Pluto Press.

———. 2003. *I am Dynamite: An Alternative Anthropology of Power*. London; New York: Routledge.

Ratto, Matt. 2009. 'Tim Ingold's Shoe'. Paper presented at SAnECH Seminars, University of Aberdeen, February.

Reed, Adam. 1999. 'Anticipating Individuals: Modes of Vision and Their Social Consequence in a Papua New Guinean Prison'. *The Journal of the Royal Anthropological Institute*. Vol. 5 (1): 43–56.

———. 2005. 'My Blog Is Me: Texts and Persons in UK Online Journal Culture (and Anthropology)'. *Ethnos*. Vol. 70 (2): 220–242.

Richards, Paul. 1993. 'Cultivation: Knowledge or Performance?' in *An Anthropological Critique of Development: The Growth of Ignorance*. Mark Hobart, ed. London; New York: Routledge.

Ricoeur, Paul. 1979. 'The Model of the Text: Meaningful Action Considered as a Text', in *Interpretive Social Science, a Reader*. Paul Rabinow and William M. Sullivan, eds. Berkeley, CA: University of California Press, pp. 73–101.

Riles, Annelise. 2000. *The Network Inside Out*. Ann Arbor: The University of Michigan Press.

Robertson, Roland. 1995. 'Glocalization: Time-Space and Homogeneity-Heterogeneity', in *Global Modernities*. Michael Featherstone, Scot Lash and Richard Robertson, eds. London: Sage.

Rodríguez, Sylvia. 2015. 'Mutuality and the Field at Home', in *Mutuality: Anthropology's Changing Terms of Engagement*. Roger Sanjek, ed. Philadelphia: University of Pennsylvania Press, pp. 45–60.

Rogister, John and Vergati, Anne. 2004. 'Introduction: Tradition Revisited'. *History and Anthropology*. Vol. 15 (3): 201–205.

Rouse, Roger. 2002. 'Mexican Migration and the Social Space of Postmodernism' in *The Anthropology of Globalization: A Reader*. Javier Inda and Renato Rosaldo, eds. Oxford: Blackwell.

Sayre, Nathan. 2012. 'The Politics of the Anthropogenic'. *Annual Review of Anthropology*. Vol. 41: 57–70.

Scarry, Elaine. 1994 [1985]. *The Body in Pain*. New York: Oxford University Press.

Schurman, Rachel and Munro, William. 2006. 'Ideas, Thinkers and Social Networks: The Process of Grievance Construction in the Anti-Genetic Engineering Movement'. *Theory and Society*. Vol. 35: 1–38.

Shapin, Steven and Schaffer, Simon. 1985. *Leviathan and the Air-Pump: Hobbes, Boyle, and the Experimental Life*. Princeton, NJ: Princeton University Press.

Shepherd, Nicole. 2002. 'Anarcho-Environmentalists: Ascetics of Late Modernity'. *Journal of Contemporary Ethnography*. Vol. 31 (2): 135–157.

Shore, Cris and Wright, Sue. 1997. 'Policy: A New Field of Anthropology', in *Anthropology of Policy: Critical Perspectives on Governance and Power*. Cris Shore and Susan Wright, eds. London; New York: Routledge.

Shore, Cris, Wright, Susan and Però, David. 2011. 'Introduction: Conceptualising Policy: Technologies of Governance and the Politics of Visibility', in Cris Shore, Susan Wright and Davide Però. eds. *Policy Worlds: Anthropology and the Analysis of Contemporary Power*. New York; Oxford: Berghahn Books.

Shotter, John. 2010. 'Movements of Feeling and Moments of Judgement: Towards an Ontological Social Constructionism'. *International Journal of Action Research*. Vol. 6 (1): 1–27.

Simpson, Bob. 2004. 'On the Impossibility of Invariant Repetition: Ritual, Tradition and Creativity Among Sri Lankan Ritual Specialists'. *History and Anthropology*. Vol. 15 (3): 301–316.

Starrett, Gregory. 1995. 'The Hexis of Interpretation: Islam and the Body in the Egyptian Popular School'. *American Ethnologist*. Vol. 16 (2).

Stone, Allucquére. 1991. 'Will the Real Body Please Stand Up? Boundary Stories About Virtual Cultures', in *Cyberspace: First Steps*. Michael Benedikt, ed. Cambridge, MA: MIT Press.

Strang, Veronica. 1997. *Uncommon Ground: Cultural Landscapes and Environmental Values*. Oxford; New York: Berg.

Strang, Veronica. 2004. *The meaning of Water*. Oxford, New York: Berg/Bloomsbury.

Strathern, Marilyn. 1987. 'The Limits of Auto-Anthropology', in *Anthropology at Home*. Anthony Jackson, ed. London; New York: Tavistock Publications.

———. 1990. 'Artefacts of History', in *Culture and History in the Pacific*. Jukka Siikala, ed. Helsinki: The Finnish Anthropological Society.

———. 1991. *Partial Connections*. New York; Toronto; Oxford: Altamira Press.

———. 1996. 'Cutting the Network'. *The Journal of the Royal Anthropological Institute*. Vol. 2: 517–535.

———. 2000. 'Abstraction and Decontextualisation: An Anthropological Comment or: E for Ethnography'. Paper presented at Virtual Society? Get Real! conference, 4–5 May. http://virtualsociety.sbs.ox.ac.uk/GRpapers/strathern.htm Accessed 9 June 2009.

SVPP working document in preparation for Abuja BGM 2006, FoE Malta Archives, Valletta.

Taussig, Michael. 1992. 'Maleficium: State Fetishism', in *The Nervous System*. Michael Taussig, ed. New York; London: Routledge.

Theodossopoulos, Dimitri. 1997. 'Turtles, Farmers and "Ecologists": The Cultural Reason Behind a Community's Resistance to Environmental Conservation'. *The Journal of Mediterranean Studies* #2. Vol. 7. Malta: University of Malta.

Throop, Jason. 2010. 'Ethical Modalities of Being and Humans Rights Discourse in YAP, FSM'. Paper presented at EASA, Maynooth, 24–28 August.

Timmer, Vanessa. 2007. 'Agility and Resilience: The Adaptive Capacity of Friends of the Earth International and Greenpeace'. PhD thesis, University of British Columbia.

Tsing, Anna. 2002. 'The Global Situation', in *The Anthropology of Globalization: A Reader*. Javier Inda and Renato Rosaldo, eds. Oxford: Blackwell.

———. 2005. *Friction: An Ethnography of Global Connection*. Oxford; Princeton, NJ: Princeton University Press.

Turkle, Sherry. 1984. *The Second Self: Computers and the Human Spirit*. New York: Simon and Schuster.

Turner, Victor. 1966. 'Communitas: Model and Process', in *The Ritual Process*. Victor Turner, ed. Ithaca, NY: Cornell.

———. 1979 'Betwixt and Between: The Liminal Period in Rites de Passage', in *Reader in Comparative Religion: An Anthropological Approach*. William A. Lessa and Evon Z. Vogt, eds. New York: Harper and Row.

Urban, Theresa. 2001. *Missão (Quase) Impossível: Aventuras e Desaventuras do Movimento Ambientalista no Brasil*. São Paulo: Peirópolis.

Urry, John. 2002. 'Mobility and Proximity'. *Sociology*. Vol. 36 (2): 255–274.

———. 2003. 'Social Network, Travel and Talk'. *British Journal of Sociology*. Vol. 54 (2): 155–175.

Van der Veer, Peter. 1989. 'The Power of Detachment: Disciplines of Body and Mind in the Ramanandi Order'. *American Ethnologist*. Vol. 16 (3): 458–470.

Vansina, Jan. 1985. *Oral Tradition as History*. London: James Currey.

Viveiros de Castro, Eduardo. 1998. 'Cosmological Deixis and Amerindian Perspectivism'. *The Journal of the Royal Anthropological Institute*. Vol. 4 (3): 469–488.

Viveiros de Castro, Eduardo, Holbraad, Martin and Axel Pedersen, Morten. 2014.' The Politics of Ontology: Anthropological Positions', in *Fieldsights—Theorizing the Contemporary, Cultural Anthropology Online*, January 13, 2014. www.culanth.org/fieldsights/462-the-politics-of-ontology-anthropological-positions

Warkentin, Craig. 2001. *Reshaping World Politics: NGOs, the Internet, and Global Civil Society*. Oxford: Rowman and Littlefield Publishers.

Washbourne, Neil. 1999. 'Beyond Iron Laws: Information Technology and Social Transformation in the Global Environmental Movement'. PhD thesis, University of Surrey.

———. 2001 'Information Technology and New Forms of Organising? Translocalism and Networks in Friends of the Earth', in *Culture and Politics in the Information Age: A New Politics?* Frank Webster, ed. London: Routledge.

Weber, Max. 2002 [1905]. *The Protestant Ethic and The Spirit of Capitalism*. London; new York: Penguin Books, translated by Peter Baehr and Gordon C. Wells.

Weiner, James. 2007. 'History, Oral History and Memoriation in Native Title', in *The Social Effects of Native Title: Recognition, Translation, Coexistence*. Benjamin R. Smith and Frances Morphy, eds. Australia: ANU E Press.

White, Richard. 1999. 'The nationalization of nature'. *Journal of American History*. Vol. 86 (3): 976–986.

Wikan, Unni. 1993. 'Beyond the Words: The Power of Resonance', in *Beyond Boundaries: Understanding, Translation and Anthropological Discourses*. Gisli Palsson, ed. Oxford; Providence: Berg.

Willerslev, Rane. 2004. 'Not Animal, Not-Animal: Hunting, Imitation, and Empathetic Knowledge Among the Siberian Yukaghirs'. *Journal of the Royal Anthropological Institute*. Vol. 10: 629–652.

Wilson, Samuel and Peterson, Leighton. 2002. 'The Anthropology of Online Communities'. *Annual Review of Anthropology*. Vol. 31: 449–467.

Wogan, Peter. 2001. 'Imagined Communities Reconsidered: Is Print-Capitalism What We Think It Is?' *Anthropological Theory*. Vol. 1: 403–418.

Woolgar, Steve. 2002. 'Five Rules of Virtuality', in Woolgar Steve (Ed), *Virtual Society?: Technology, Cyberbole, Reality*. Oxford; New York: Oxford Unviersity Press.

Wright, Sue. 1994. 'Culture in Anthropology and Organizational Studies', in *Anthropology of Organizations*. Sue Wright, ed. London; New York: Routledge.

———. 2008. 'Governance as a Regime of Discipline', in *Exploring Regimes of Discipline: The Dynamics of Restraint*. Noel Dyck, ed. London; New York: Berghahn Books.

Wulff, Helena. 2000. 'Methods for a Multi-Locale Study of Ballet as a Career', in *Constructing the Field*. Vered Amit, ed. London: Routledge.

Wulff, Helena. 2003. 'Steps on Screen: Technoscapes, Visualization and Globalization in Dance', in *New Technologies at Work: People, Screens and Social Virtuality*. Christina Garsten and Helena Wulff, eds. USA; UK: Berg.

Yarrow, Thomas. 2006. 'Developing Knowledge: Knowledge, Relationships and Ideology Amongst Ghanaian Development Workers'. PhD thesis, University of Cambridge.

———. 2008. 'Life/History: Personal Narratives of Development in Ghana'. *Africa*. Vol. 78: 334–357.

———. 2011. *Development Beyond Politics: Aid, Activism and NGOs in Ghana*. Basingstoke: Palgrave Macmillan.

Zammit, David. 1998. *Laws and Stories: An Ethnographic Study of Maltese Legal Representation*, PhD thesis, University of Durham.

Zimmerman Umble, Diane. 1992. 'The Amish and the Telephone: Resistance and Reconstruction', in *Consuming Technologies: Media and Information in Domestic Spaces*. Roger Silverstone and Eric Hirsch, eds. London: Routledge.

Index